T0336121

INTRODUCTION TO FOLIATIONS AND LIE GROUPOIDS

This book gives a quick introduction to the theory of foliations, Lie groupoids, and Lie algebroids. An important feature is the emphasis on the interplay between these concepts; Lie groupoids form an indispensable tool for the study of the transverse structure of foliations as well as their non-commutative geometry, while the theory of foliations has immediate applications to the Lie theory of groupoids and their infinitesimal algebroids.

This book starts with a detailed presentation of the main classical theorems in the theory of foliations, then proceeds to Molino's theory, Lie groupoids, constructing the holonomy groupoid of a foliation, and finally Lie algebroids. Among other things, the authors discuss to what extent Lie's theory for Lie groups and Lie algebras holds in the more general context of groupoids and algebroids. Based on the authors' extensive teaching experience, this book contains numerous examples and exercises, making it ideal for graduate students and their instructors.

CAMBRIDGE STUDIES IN ADVANCED MATHEMATICS

Editorial Board:

B. Bollobás, W. Fulton, A. Katok, F. Kirwan, P. Sarnak, B. Simon

Already published

17 W. Dicks & M. Dunwoody *Groups acting on graphs*
18 L.J. Corwin & F.P. Greenleaf *Representations of nilpotent Lie groups and their applications*
19 R. Fritsch & R. Piccinini *Cellular structures in topology*
20 H. Klingen *Introductory lectures on Siegel modular forms*
21 P. Koosis *The logarithmic integral II*
22 M.J. Collins *Representations and characters of finite groups*
24 H. Kunita *Stochastic flows and stochastic differential equations*
25 P. Wojtaszczyk *Banach spaces for analysis*
26 J.E. Gilbert & M.A.M. Murray *Clifford algebras and Dirac operators in harmonic analysis*
27 A. Fröhlich & M.J. Taylor *Algebraic number theory*
28 K. Goebel & W.A. Kirk *Topics in metric fixed point theory*
29 J.F. Humphreys *Reflection groups and Coxeter groups*
30 D.J. Benson *Representations and cohomology I*
31 D.J. Benson *Representations and cohomology II*
32 C. Allday & V. Puppe *Cohomological methods in transformation groups*
33 C. Soulé et al. *Lectures on Arakelov geometry*
34 A. Ambrosetti & G. Prodi *A primer of nonlinear analysis*
35 J. Palis & F. Takens *Hyperbolicity, stability and chaos at homoclinic bifurcations*
37 Y. Meyer *Wavelets and aperators I*
38 C. Weibel *An introduction to homological algebra*
39 W. Bruns & J. Herzog *Cohen–Macaulay rings*
40 V. Snaith *Explicit Brauer induction*
41 G. Laumon *Cohomology of Drinfeld modular varieties I*
42 E.B. Davies *Spectral theory and differential operators*
43 J. Diestel, H. Jarchow, & A. Tonge *Absolutely summing operators*
44 P. Mattila *Geometry of sets and measures in Euclidean spaces*
45 R. Pinsky *Positive harmonic functions and diffusion*
46 G. Tenenbaum *Introduction to analytic and probabilistic number theory*
47 C. Peskine *An algebraic introduction to complex projective geometry*
48 Y. Meyer & R. Coifman *Wavelets*
49 R. Stanley *Enumerative combinatorics I*
50 I. Porteous *Clifford algebras and the classical groups*
51 M. Audin *Spinning tops*
52 V. Jurdjevic *Geometric control theory*
53 H. Volklein *Groups as Galois groups*
54 J. Le Potier *Lectures on vector bundles*
55 D. Bump *Automorphic forms and representations*
56 G. Laumon *Cohomology of Drinfeld modular varieties II*
57 D.M. Clark & B.A. Davey *Natural dualities for the working algebraist*
58 J. McCleary *A user's guide to spectral sequences II*
59 P. Taylor *Practical foundations of mathematics*
60 M.P. Brodmann & R.Y. Sharp *Local cohomology*
61 J.D. Dixon et al. *Analytic pro-p groups*
62 R. Stanley *Enumerative combinatorics II*
63 R.M. Dudley *Uniform central limit theorems*
64 J. Jost & X. Li-Jost *Calculus of variations*
65 A.J. Berrick & M.E. Keating *An introduction to rings and modules*
66 S. Morosawa *Holomorphic dynamics*
67 A.J. Berrick & M.E. Keating *Categories and modules with K-theory in view*
68 K. Sato *Levy processes and infinitely divisible distributions*
69 H. Hida *Modular forms and Galois cohomology*
70 R. Iorio & V. Iorio *Fourier analysis and partial differential equations*
71 R. Blei *Analysis in integer and fractional dimensions*
72 F. Borceaux & G. Janelidze *Galois theories*
73 B. Bollobás *Random graphs*
74 R.M. Dudley *Real analysis and probability*
75 T. Sheil-Small *Complex polynomials*
76 C. Voisin *Hodge theory and complex algebraic geometry, I*
77 C. Voisin *Hodge theory and complex algebraic geometry, II*
78 V. Paulsen *Completely bounded maps and operator algebras*
82 G. Tourlakis *Lectures in logic and set theory I*
83 G. Tourlakis *Lectures in logic and set theory II*
84 R. Bailey *Association schemes*

INTRODUCTION TO
FOLIATIONS AND
LIE GROUPOIDS

I. MOERDIJK AND J. MRČUN

CAMBRIDGE
UNIVERSITY PRESS

CAMBRIDGE UNIVERSITY PRESS
Cambridge, New York, Melbourne, Madrid, Cape Town, Singapore, São Paulo

Cambridge University Press
The Edinburgh Building, Cambridge CB2 8RU, UK

Published in the United States of America by Cambridge University Press, New York

www.cambridge.org
Information on this title: www.cambridge.org/9780521831970

© Cambridge University Press 2003

This publication is in copyright. Subject to statutory exception
and to the provisions of relevant collective licensing agreements,
no reproduction of any part may take place without the written
permission of Cambridge University Press.

First published 2003
Reprinted 2005

A catalogue record for this publication is available from the British Library

Library of Congress Cataloguing in Publication data

Moerdijk, Izak.
Introduction to foliations and Lie groupoids / I. Moerdijk and J. Mrčun.
p. cm. – (Cambridge studies in advanced mathematics; 91)
Includes bibliographical references and index.
ISBN 0 521 83197 0 (hardback)
1. Foliations (Mathematics) 2. Lie groupoids. I. Mrčun, J., 1966– II. Title. III. Series.
QA613.62.M64 2003
514'.72 – dc21 2003046172

ISBN 978-0-521-83197-0 hardback

Transferred to digital printing 2007

Contents

Preface

The purpose of this book is to give a quick introduction to the theory of foliations, as well as to Lie groupoids and their infinitesimal version – Lie algebroids. The book is written for students who are familiar with the basic concepts of differential geometry, and all the results presented in this book are proved in detail.

The topics in this book have been chosen so as to emphasize the relations between foliations, Lie groupoids and Lie algebroids. Lie groupoids form the main tool for the study of the 'transversal structure' (the space of leaves) of a foliation, by means of its holonomy groupoid. Foliations are also a special kind of Lie algebroids. At the same time, the elementary theory of foliations is a very useful tool in studying Lie groupoids and Lie algebroids.

In Chapter 1 we present the basic definitions, examples and constructions of foliations. Chapter 2 introduces the notion of *holonomy*, which plays a central role in this book. The Reeb stability theorems are discussed, as well as Riemannian foliations and their holonomy. This chapter also contains an introduction to the theory of orbifolds (or V-manifolds). Orbifolds provide a language to describe the richer structure of the space of leaves of certain foliations; e.g. the space of leaves of a Riemannian foliation is often an orbifold.

In Chapter 3 we present two classical milestones of the theory of foliations in codimension 1, namely the theorems of Haefliger and Novikov, with detailed proofs. Although the proofs make essential use of the notion of holonomy, this chapter is somewhat independent of the rest of the book (see the figure). However, it should be pointed out that there are proofs of Haefliger's theorem which use the transverse structure and Lie groupoids, see e.g. Jekel (1976) or Van Est (1984).

In Chapter 4, we discuss homogeneous and transversely parallelizable foliations, as well as Lie foliations, culminating in Molino's structure theorem

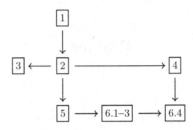

Interdependence of the chapters

for Riemannian foliations. This chapter also provides an essential link to the integrability theory of Lie groupoids and Lie algebroids.

In Chapter 5, we introduce the notion of Lie groupoid. The fine structure of the space of leaves of a foliation can be modelled by its holonomy and monodromy groupoids, and these provide some of the main examples of Lie groupoids. These Lie groupoids play an important role in the study of foliations from the point of view of non-commutative geometry as well; see Connes (1994). Orbifolds can also be viewed as Lie groupoids; in fact, they are shown to be essentially equivalent to a special class of Lie groupoids.

The infinitesimal part of a Lie groupoid gives rise to the structure of a Lie algebroid, similarly to the case of Lie groups and Lie algebras. In Chapter 6, we introduce these structures, and examine to what extent the correspondence between Lie groups and Lie algebras ('Lie theory') extends to Lie groupoids and Lie algebroids. Here we make essential use of elementary foliation theory, e.g. to construct the simply connected cover of a Lie groupoid, and to establish the correspondence of maps between Lie groupoids and maps between their Lie algebroids. Transversely parallelizable foliations from Chapter 4 provide natural examples of Lie algebroids which are not 'integrable', i.e. are not the infinitesimal parts of Lie groupoids.

This small book came into existence over a relatively long period of time. Chapters 1–3 are based on the notes of part of a course on foliations given at Utrecht University in 1995 and several subsequent years, and at the University of Ljubljana in 1997 and 1998. Chapter 4 was added later, in 1999. The last two chapters, on Lie groupoids and Lie algebroids, have been written more recently (in 2000) in this form, although much of this material had been presented by both of us in many earlier lectures and research papers.

Over the years, we have been influenced and helped by discussions with many friends and colleagues, and it would be impossible to thank them all here. However, we do wish to acknowledge our gratitude to A. Haefliger, who has been very encouraging, while it is obvious from the text that we owe a lot to his work. We are also much indebted to the late W.T. van Est who first got us interested in foliations, and to K.C.H. Mackenzie for many helpful discussions about Lie algebroids.

At a different level, we would like to thank the Dutch Science Foundation (NWO) and the Slovenian Ministry of Science (MŠZŠ grant J1-3148) for their support, as well as our home institutions for their support and hospitality, which made mutual visits possible and pleasant. Finally, we would like to thank the staff of Cambridge University Press for their help during the final stages of the preparation of this book.

Prerequisites

In this book we presuppose some familiarity with the basic notions of differential topology and geometry. Good references are e.g. Guillemin–Pollack (1974) and Bott–Tu (1982). We shall list some of these notions, partly to fix the notations.

Recall that a *smooth manifold* (or a C^∞-manifold) of dimension n (where $n = 0, 1, \ldots$) is a second-countable Hausdorff space M, together with a maximal *atlas* of open embeddings (*charts*)

$$(\varphi_i \colon U_i \longrightarrow \mathbb{R}^n)_{i \in I}$$

of open subsets $U_i \subset M$ into $\mathbb{R}^n$, such that $M = \bigcup_{i \in I} U_i$ and the *change-of-charts homeomorphisms*

$$\varphi_{ij} = \varphi_i \circ (\varphi_j|_{U_i \cap U_j})^{-1} \colon \varphi_j(U_i \cap U_j) \longrightarrow \varphi_i(U_i \cap U_j)$$

are smooth, for any $i, j \in I$. Note that these satisfy the cocycle condition $\varphi_{ij}(\varphi_{jk}(x)) = \varphi_{ik}(x)$, $x \in \varphi_k(U_i \cap U_j \cap U_k)$. There is an associated notion of a *smooth* map between smooth manifolds. Any smooth manifold (Hausdorff and second-countable) is paracompact, which is sufficient for the existence of partitions of unity.

The notions of (maximal) atlas and of smooth map also make sense if M is any topological space, not necessarily second countable or Hausdorff. We refer to such a space with a maximal atlas as a *non-Hausdorff manifold* or a *non-second-countable manifold*. There are many more non-Hausdorff manifolds than the usual Hausdorff ones, even in dimension 1 (see Haefliger–Reeb (1957)). We shall have occasion to consider such non-Hausdorff manifolds later in this book.

The reader should be familiar with the notion of the *tangent bundle* $T(M)$ of M, which is a vector bundle over M of rank n, where n is the dimension of the manifold M. The *tangent space* $T_x(M)$ of M at

$x \in M$ is the fibre of $T(M)$ over x. The (smooth) sections of the tangent bundle $T(M)$ are the *vector fields* on M. The $C^\infty(M)$-module $\mathfrak{X}(M)$ of all vector fields on M is a Lie algebra, and the Lie bracket on $\mathfrak{X}(M)$ satisfies the Leibniz identity

$$[X, fY] = f[X, Y] + X(f)Y$$

for all $X, Y \in \mathfrak{X}(M)$ and $f \in C^\infty(M)$.

Also, we have the space $\Omega^k(M)$ of *differential k-forms* on M, for any $k = 0, 1, \ldots, n$, with *exterior differentiation* $d\colon \Omega^k(M) \to \Omega^{k+1}(M)$ and *exterior product* $\wedge\colon \Omega^k(M) \otimes \Omega^l(M) \to \Omega^{k+l}(M)$. With this, $\Omega(M) = \bigoplus_{k=0}^{n} \Omega^k(M)$ becomes a differential graded algebra, which is commutative (in the graded sense). The cohomology of $(\Omega(M), d)$ is called the *de Rham cohomology* of M, and denoted by

$$H_{\mathrm{dR}}(M) = \bigoplus_{k=0}^{n} H_{\mathrm{dR}}^k(M) \, .$$

A smooth map $f\colon M \to N$ between smooth manifolds has a *derivative* $df\colon T(M) \to T(N)$, which is a bundle map over f. The derivative of f at $x \in M$ is the restriction of df to the corresponding tangent spaces over x and $f(x)$, and denoted by $(df)_x\colon T_x(M) \to T_{f(x)}(N)$. The map f is an *immersion* if each $(df)_x$ is injective, and a *submersion* if each $(df)_x$ is surjective. These have canonical local forms on a small neighbourhood of $x \in M$:

(i) If f is an immersion, there exist open neighbourhoods $U \subset M$ of x and $V \subset N$ of $f(x)$ with $f(U) \subset V$ and diffeomorphisms $\varphi\colon U \to \mathbb{R}^n$ and $\psi\colon V \to \mathbb{R}^p$ such that

$$(\psi \circ f \circ \varphi^{-1})(y) = (y, 0)$$

with respect to the decomposition $\mathbb{R}^p = \mathbb{R}^n \times \mathbb{R}^{p-n}$.

(ii) If f is a submersion, there exist open neighbourhoods $U \subset M$ of x and $V \subset N$ of $f(x)$ with $f(U) = V$ and diffeomorphisms $\varphi\colon U \to \mathbb{R}^n$ and $\psi\colon V \to \mathbb{R}^p$ such that

$$(\psi \circ f \circ \varphi^{-1})(y, z) = y$$

with respect to the decomposition $\mathbb{R}^n = \mathbb{R}^p \times \mathbb{R}^{n-p}$.

A smooth map $f\colon M \to N$ is a *diffeomorphism* if it is a bijection and has a smooth inverse. The map f is a *local diffeomorphism* (or *étale map*) if $(df)_x$ is an isomorphism for any $x \in M$. Any bijective local diffeomorphism is a diffeomorphism.

A smooth map $g \colon K \to N$ is said to be an *embedding* if it is an immersion and a topological embedding. This makes K a *submanifold* of N, and $T(K)$ a subbundle of $T(N)$.

If K is a submanifold of N and $f \colon M \to N$ a smooth map, one says that f is *transversal* to K if $(df)_x(T_x(M)) + T_{f(x)}(K) = T_{f(x)}(N)$ for every $x \in f^{-1}(K)$.

For every submanifold K of N there exists an open neighbourhood $U \subset N$ of K which has the structure of a vector bundle over K, with the inclusion $K \hookrightarrow U$ corresponding to the zero section. In particular, the projection $U \to K$ of this bundle is a retraction. Such a U is called a *tubular neighbourhood* of K.

Recall that, on a vector bundle E of rank n over a manifold M, one can always choose a *Riemannian structure* (by using partitions of unity). A *Riemannian metric* on M is a Riemannian structure on $T(M)$. The structure group of E can be reduced to $O(n)$. The bundle E is called *orientable* if its structure group can be reduced to $SO(n)$. An *orientation* of an orientable vector bundle E is an equivalence class of oriented trivializations of E.

1
Foliations

Intuitively speaking, a foliation of a manifold M is a decomposition of M into immersed submanifolds, the leaves of the foliation. These leaves are required to be of the same dimension, and to fit together nicely.

Such foliations of manifolds occur naturally in various geometric contexts, for example as solutions of differential equations and integrable systems, and in symplectic geometry. In fact, the concept of a foliation first appeared explicitly in the work of Ehresmann and Reeb, motivated by the question of existence of completely integrable vector fields on three-dimensional manifolds. The theory of foliations has now become a rich and exciting geometric subject by itself, as illustrated be the famous results of Reeb (1952), Haefliger (1956), Novikov (1964), Thurston (1974), Molino (1988), Connes (1994) and many others.

We start this book by describing various equivalent ways of defining foliations. A foliation on a manifold M can be given by a suitable foliation atlas on M, by an integrable subbundle of the tangent bundle of M, or by a locally trivial differential ideal. The equivalence of all these descriptions is a consequence of the Frobenius integrability theorem. We will give several elementary examples of foliations. The simplest example of a foliation on a manifold M is probably the one given by the level sets of a submersion $M \to N$. In general, a foliation on M is a decomposition of M into leaves which is *locally* given by the fibres of a submersion.

In this chapter we also discuss some first properties of foliations, for instance the property of being orientable or transversely orientable. We show that a transversely orientable foliation of codimension 1 on a manifold M is given by the kernel of a differential 1-form on M, and that this form gives rise to the so-called Godbillon–Vey class. This is a class of degree 3 in the de Rham cohomology of M, which depends only on the foliation and not on the choice of the specific 1-form. Furthermore, we

4

discuss here several basic methods for constructing foliations. These include the product and pull-back of foliations, the formation of foliations on quotient manifolds, the construction of foliations by 'suspending' a diffeomorphism or a group of diffeomorphisms, and foliations associated to actions of Lie groups.

1.1 Definition and first examples

Let M be a smooth manifold of dimension n. A *foliation atlas* of codimension q of M (where $0 \leq q \leq n$) is an atlas

$$(\varphi_i \colon U_i \longrightarrow \mathbb{R}^n = \mathbb{R}^{n-q} \times \mathbb{R}^q)_{i \in I}$$

of M for which the change-of-charts diffeomorphisms φ_{ij} are locally of the form

$$\varphi_{ij}(x, y) = (g_{ij}(x, y), h_{ij}(y))$$

with respect to the decomposition $\mathbb{R}^n = \mathbb{R}^{n-q} \times \mathbb{R}^q$. The charts of a foliation atlas are called the *foliation charts*. Thus each U_i is divided into *plaques*, which are the connected components of the submanifolds $\varphi_i^{-1}(\mathbb{R}^{n-q} \times \{y\})$, $y \in \mathbb{R}^q$, and the change-of-charts diffeomorphisms preserve this division (Figure 1.1). The plaques globally amalgamate

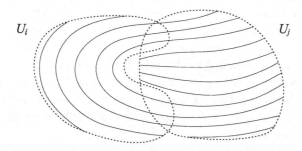

Fig. 1.1. Two foliation charts

into *leaves*, which are smooth manifolds of dimension $n - q$ injectively immersed into M. In other words, two points $x, y \in M$ lie on the same leaf if there exist a sequence of foliation charts $U_1, \ldots, U_k$ and a sequence of points $x = p_0, p_1, \ldots, p_k = y$ such that p_{j-1} and p_j lie on the same plaque in U_j, for any $1 \leq j \leq k$.

A *foliation* of *codimension* q of M is a maximal foliation atlas of M of codimension q. Each foliation atlas determines a foliation, since it is included in a unique maximal foliation atlas. Two foliation atlases define the same foliation of M precisely if they induce the same partition of M into leaves. A (smooth) *foliated manifold* is a pair $(M, \mathcal{F})$, where M is a smooth manifold and $\mathcal{F}$ a foliation of M. The *space of leaves* $M/\mathcal{F}$ of a foliated manifold $(M, \mathcal{F})$ is the quotient space of M, obtained by identifying two points of M if they lie on the same leaf of $\mathcal{F}$. The dimension of $\mathcal{F}$ is $n - q$. A (smooth) map between foliated manifolds $f \colon (M, \mathcal{F}) \to (M', \mathcal{F}')$ is a (smooth) map $f \colon M \to N$ which preserves the foliation structure, i.e. which maps leaves of $\mathcal{F}$ into the leaves of $\mathcal{F}'$.

This is the first definition of a foliation. Instead of smooth foliations one can of course consider C^r-foliations, for any $r \in \{0, 1, \ldots, \infty\}$, or (real) analytic foliations. Standard references are Bott (1972), Hector–Hirsch (1981, 1983), Camacho–Neto (1985), Molino (1988) and Tondeur (1988). In the next section we will give several equivalent definitions: in terms of a Haefliger cocycle, in terms of an integrable subbundle of $T(M)$, and in terms of a differential ideal in $\Omega(M)$. But first we give some examples.

Examples 1.1 (1) The space $\mathbb{R}^n$ admits the *trivial* foliation of codimension q, for which the atlas consists of only one chart id: $\mathbb{R}^n \to \mathbb{R}^{n-q} \times \mathbb{R}^q$. Of course, any linear bijection $A \colon \mathbb{R}^n \to \mathbb{R}^{n-q} \times \mathbb{R}^q$ determines another one whose leaves are the affine subspaces $A^{-1}(\mathbb{R}^{n-q} \times \{y\})$.

(2) Any submersion $f \colon M \to N$ defines a foliation $\mathcal{F}(f)$ of M whose leaves are the connected components of the fibres of f. The codimension of $\mathcal{F}(f)$ is equal to the dimension of N. An atlas representing $\mathcal{F}(f)$ is derived from the canonical local form for the submersion f. Foliations associated to the submersions are also called *simple* foliations. The foliations associated to submersions with connected fibres are called *strictly simple*. A simple foliation is strictly simple precisely when its space of leaves is Hausdorff.

(3) (Kronecker foliation of the torus) Let a be an irrational real number, and consider the submersion $s \colon \mathbb{R}^2 \to \mathbb{R}$ given by $s(x, y) = x - ay$. By (2) we have the foliation $\mathcal{F}(s)$ of $\mathbb{R}^n$. Let $f \colon \mathbb{R}^2 \to T^2 = S^1 \times S^1$ be the standard covering projection of the torus, i.e. $f(x, y) = (e^{2\pi i x}, e^{2\pi i y})$. The foliation $\mathcal{F}(s)$ induces a foliation $\mathcal{F}$ of T^2: if φ is a foliation chart for $\mathcal{F}(s)$ such that $f|_{\mathrm{dom}\varphi}$ is injective, then $\varphi \circ (f|_{\mathrm{dom}\varphi})^{-1}$ is a foliation chart for $\mathcal{F}$. Any leaf of $\mathcal{F}$ is diffeomorphic to $\mathbb{R}$, and is dense in T^2 (Figure 1.2).

(4) (Foliation of the Möbius band) Let $f \colon \mathbb{R}^2 \to M$ be the standard covering projection of the (open) Möbius band: $f(x, y) = f(x', y')$

Fig. 1.2. Kronecker foliation of the torus

precisely if $x' - x \in \mathbb{Z}$ and $y' = (-1)^{x'-x}y$. The trivial foliation of codimension 1 of $\mathbb{R}^2$ induces a foliation $\mathcal{F}$ of M, in the same way as in (3). All the leaves of $\mathcal{F}$ are diffeomorphic to S^1, and they are wrapping around M twice, except for the 'middle' one: this one goes around only once (Figure 1.3).

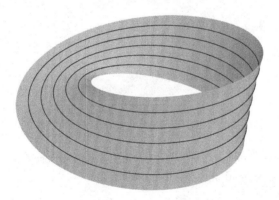

Fig. 1.3. Foliation of the Möbius band

(5) (The Reeb foliation of the solid torus and of S^3) One can also define the notion of a foliation of a manifold with boundary in the obvious way; however, one usually assumes that the leaves of such a foliation behave well near the boundary, by requiring either that they are transversal to the boundary, or that the connected components of the boundary are leaves. An example of the last sort is the Reeb foliation of the solid torus, which is given as follows.

Consider the unit disk $D = \{z \mid z \in \mathbb{C}, |z| \leq 1\}$, and define a submersion $f\colon \mathrm{Int}(D) \times \mathbb{R} \to \mathbb{R}$ by

$$f(z,x) = \mathrm{e}^{\overline{\cdot\,-|z|^{\cdot}}} - x \ .$$

So we have the foliation $\mathcal{F}(f)$ of $\mathrm{Int}(D) \times \mathbb{R}$, which can be extended to a foliation of the cylinder $D \times \mathbb{R}$ by adding one new leaf: the boundary $S^1 \times \mathbb{R}$. Now $D \times \mathbb{R}$ is a covering space of the solid torus $X = D \times S^1$ in the canonical way, and the foliation of $D \times \mathbb{R}$ induces a foliation of the solid torus. We will denote this foliation by $\mathcal{R}$. The boundary torus of this solid torus is a leaf of $\mathcal{R}$. Any other leaf of $\mathcal{R}$ is diffeomorphic to $\mathbb{R}^2$, and has the boundary leaf as its set of adherence points in X. The *Reeb foliation* of X is any foliation $\mathcal{F}$ of X of codimension 1 for which there exists a homeomorphism of X which maps the leaves of $\mathcal{F}$ onto the leaves of $\mathcal{R}$ (Figure 1.4).

Fig. 1.4. The Reeb foliation of the solid torus

The three-dimensional sphere S^3 can be decomposed into two solid tori glued together along their boundaries, i.e.

$$S^3 \cong X \cup_{\partial X} X \ .$$

Since ∂X is a leaf of the Reeb foliation of X, we can glue the Reeb foliations of both copies of X along ∂X as well. This can be done so that the obtained foliation of S^3 is smooth. This foliation has a unique compact leaf and is called the *Reeb foliation* of S^3.

Exercise 1.2 Describe in each of these examples explicitly the space of leaves of the foliation. (You will see that this space often has a very poor structure. Much of foliation theory is concerned with the study of 'better models' for the leaf space.)

1.2 Alternative definitions of foliations

A foliation $\mathcal{F}$ of a manifold M can be equivalently described in the following ways (here n is the dimension of M and q the codimension of $\mathcal{F}$).

(i) By a foliation atlas $(\varphi_i : U_i \to \mathbb{R}^{n-q} \times \mathbb{R}^q)$ of M for which the change-of-charts diffeomorphisms φ_{ij} are *globally* of the form $\varphi_{ij}(x, y) = (g_{ij}(x, y), h_{ij}(y))$ with respect to the decomposition $\mathbb{R}^n = \mathbb{R}^{n-q} \times \mathbb{R}^q$.

(ii) By an open cover (U_i) of M with submersions $s_i \colon U_i \to \mathbb{R}^q$ such that there are diffeomorphisms (necessarily unique)

$$\gamma_{ij} \colon s_j(U_i \cap U_j) \longrightarrow s_i(U_i \cap U_j)$$

with $\gamma_{ij} \circ s_j|_{U_i \cap U_j} = s_i|_{U_i \cap U_j}$. (The diffeomorphisms γ_{ij} satisfy the cocycle condition $\gamma_{ij} \circ \gamma_{jk} = \gamma_{ik}$. This cocycle is called the *Haefliger cocycle* representing $\mathcal{F}$.)

(iii) By an *integrable subbundle* E of $T(M)$ of rank $n - q$. (Here integrable (or involutive) means that E is closed under the Lie bracket, i.e. if $X, Y \in \mathfrak{X}(M)$ are sections of E, then the vector field $[X, Y]$ is also a section of E.)

(iv) By a locally trivial *differential* (graded) *ideal* $\mathcal{J} = \bigoplus_{k=1}^{n} \mathcal{J}^k$ of rank q in the differential graded algebra $\Omega(M)$. (An ideal $\mathcal{J}$ is locally trivial of rank q if any point of M has an open neighbourhood U such that $\mathcal{J}|_U$ is the ideal in $\Omega(M)|_U$ generated by q linearly independent 1-forms. An ideal $\mathcal{J}$ is differential if $d\mathcal{J} \subset \mathcal{J}$.)

Before we go into details of why these descriptions of the concept of foliation are equivalent, we should point out that the bundle E of (iii) consists of tangent vectors to M which are tangent to the leaves, while a differential k-form is in the ideal $\mathcal{J}$ of (iv) if it vanishes on any k-tuple of vectors which are all tangent to the leaves.

Ad (i): Any foliation atlas $(\varphi_i \colon U_i \to \mathbb{R}^{n-q} \times \mathbb{R}^q)$ of $\mathcal{F}$ has a refinement which satisfies the condition in (i). To see this, we may first assume that (U_i) is a locally finite cover of M. Next, we may find a locally finite refinement (V_k) of (U_i) such that $V_k \cup V_l$ is contained in some U_i for any non-disjoint V_k and V_l. As any V_k is contained in a U_{i_k}, we may take $\psi_k = \varphi_{i_k}|_{V_k}$. Further we may choose each V_k so small that for any $U_j \supset V_k$, the change-of-charts diffeomorphism $\varphi_j \circ \psi_k^{-1}$ is globally of the form $(g_{jk}(x, y), h_{jk}(y))$, and that h_{jk} is an embedding. This refinement (ψ_k) of (φ_i) is a foliation atlas of M which satisfies the condition in (i).

Ad (ii): If (U_i, s_i, γ_{ij}) is a Haefliger cocycle on M, choose an atlas $(\varphi_k \colon V_k \to \mathbb{R}^n)$ so that each V_k is a subset of an U_{i_k} and φ_k renders s_{i_k} in the normal form for a submersion: it is surjective, and there exists a diffeomorphism $\psi_k \colon s_{i_k}(V_k) \to \mathbb{R}^q$ such that $\psi_k \circ s_{i_k} = \mathrm{pr}_2 \circ \varphi_k$. This is a foliation atlas of the form in (i): if $(x, y) \in \varphi_k(V_k \cap V_l) \subset \mathbb{R}^{n-q} \times \mathbb{R}^q$, we have

$$
\begin{aligned}
(\mathrm{pr}_2 \circ \varphi_l \circ \varphi_k^{-1})(x, y) &= (\psi_l \circ s_{i_l} \circ \varphi_k^{-1})(x, y) \\
&= (\psi_l \circ \gamma_{i_l i_k} \circ s_{i_k} \circ \varphi_k^{-1})(x, y) \\
&= (\psi_l \circ \gamma_{i_l i_k} \circ \psi_k)(y) \,.
\end{aligned}
$$

Conversely, if $(\varphi_i \colon U_i \to \mathbb{R}^{n-q} \times \mathbb{R}^q)$ is a foliation atlas of the form in (i), take $s_i = \mathrm{pr}_2 \circ \varphi_i$ and $\gamma_{ij} = h_{ij}$. This gives a Haefliger cocycle on M which represents the same foliation.

Ad (iii): Let us assume that the foliation is given by a foliation atlas $(\varphi_i \colon U_i \to \mathbb{R}^{n-q} \times \mathbb{R}^q)$. Define a subbundle E of $T(M)$ locally over U_i by

$$
E|_{U_i} = \mathrm{Ker}(d(\mathrm{pr}_2 \circ \varphi_i)) \,,
$$

i.e. by the kernel of the $\mathbb{R}^q$-valued 1-form $\alpha = d(\mathrm{pr}_2 \circ \varphi_i)$. For any such a 1-form and any vector fields X, Y on U_i we have $2d\alpha(X, Y) = X(\alpha(Y)) - Y(\alpha(X)) - \alpha([X, Y])$. Since our α is closed, it follows that

$$
\alpha([X, Y]) = X(\alpha(Y)) - Y(\alpha(X)) \,.
$$

Using this it is clear that E is an integrable subbundle of $T(M)$ of codimension q.

The bundle E is uniquely determined by the foliation $\mathcal{F}$: a tangent vector $\xi \in T_x(M)$ is in E precisely if ξ is tangent to the leaf of L through x. The bundle E is called the *tangent bundle* of $\mathcal{F}$, and is often denoted by $T(\mathcal{F})$. A section of $T(\mathcal{F})$ is called a vector field *tangent* to $\mathcal{F}$. The Lie algebra $\Gamma(T(\mathcal{F}))$ of sections of $T(\mathcal{F})$ will also be denoted by $\mathfrak{X}(\mathcal{F})$.

Conversely, an integrable subbundle E of codimension q of $T(M)$ can be locally integrated (Frobenius theorem, see Appendix of Camacho–Neto (1985)): for any point $x \in M$ there exist an open neighbourhood $U \subset M$ and a diffeomorphism $\varphi \colon U \to \mathbb{R}^{n-q} \times \mathbb{R}^q$ such that $E|_U = \mathrm{Ker}(d(\mathrm{pr}_2 \circ \varphi))$. By using these kinds of diffeomorphisms as foliation charts, one obtains a foliation atlas of the foliation.

Ad (iv): For any subbundle E of $T(M)$, define the (graded) ideal $\mathcal{J} = \bigoplus_{k=1}^n \mathcal{J}^k$ in $\Omega(M)$ as follows: for $\omega \in \Omega^k(M)$,

$$
\omega \in \mathcal{J}^k \text{ if and only if}
$$

$\omega(X_1, \ldots, X_k) = 0$ for any sections $X_1, \ldots, X_k$ of E.

Note that $\mathcal{J}$ is locally trivial of rank q, i.e. it is locally generated by q linearly independent 1-forms: Choose a local frame $X_1, \ldots, X_n$ of $T(M)|_U$ such that $X_1, \ldots, X_{n-q}$ form a frame of $E|_U$. There is the dual frame of differential 1-forms $\omega_1, \ldots, \omega_n$ of $T(M)^*|_U$, and the linearly independent 1-forms $\omega_{n-q+1}, \ldots, \omega_n$ clearly generate the ideal $\mathcal{J}$. Conversely, any locally trivial ideal $\mathcal{J}$ of rank q determines a subbundle E of $T(M)$ of rank $n - q$, by the formula above (for $k = 1$).

We claim that under this correspondence,

$\mathcal{J}$ is differential if and only if E is integrable.

In fact, this is immediate from the definition of the exterior derivative:

$$d\omega(X_0, \ldots, X_k) = \frac{1}{k+1} \sum_{0 \le i \le k} (-1)^i X_i(\omega(X_0, \ldots, \hat{X}_i, \ldots, X_k))$$

$$+ \frac{1}{k+1} \sum_{0 \le j < l \le k} (-1)^{j+l} \omega([X_j, X_l], X_0, \ldots, \hat{X}_j, \ldots, \hat{X}_l, \ldots, X_k) .$$

REMARKS. (1) Let $\mathcal{J}$ be a locally trivial ideal of rank q in $\Omega(M)$. If $\mathcal{J}$ is differential and locally (over U) generated by $\omega_1, \ldots, \omega_q$, then

$$d\omega_i = \sum_{j=1}^{q} \alpha_{ij} \wedge \omega_j$$

for some $\alpha_{ij} \in \Omega^1(M)|_U$. In particular,

$$d\omega_i \wedge \omega_1 \wedge \cdots \wedge \omega_q = 0 .$$

Conversely, if we have an open cover (U_l) of M such that for any l the restriction $\mathcal{J}|_{U_l}$ is generated by linearly independent 1-forms $\omega_1^l, \ldots, \omega_q^l$ satisfying

$$d\omega_i^l \wedge \omega_1^l \wedge \cdots \wedge \omega_q^l = 0$$

for any i, then $\mathcal{J}$ is differentiable. Indeed, this implies that

$$d\omega_i^l = \sum_{j=1}^{q} \alpha_{ij}^l \wedge \omega_j^l$$

for some $\alpha_{ij}^l \in \Omega^1(M)|_{U_l}$; to see this, one should locally complete $\omega_1^l, \ldots, \omega_q^l$ to a frame and compute α_{ij}^l locally, and finally obtain α_{ij}^l on U_l using partition of unity (exercise: fill in the details).

(2) A one-dimensional subbundle E of $T(M)$ (i.e. a *line field*) is clearly integrable, hence any line field on M defines a foliation of M of codimension $n - 1$.

(3) If ω is a nowhere vanishing 1-form on M, it defines a foliation of codimension 1 of M precisely if it is *integrable*, i.e. if

$$d\omega \wedge \omega = 0 \ .$$

Note that if $\dim M = 2$ then any 1-form ω on M is integrable.

In particular, any closed 1-form on M is integrable. For example, if $\omega = df$ for a smooth map $f \colon M \to \mathbb{R}$ without critical points, this gives exactly the foliation given by the submersion f.

Note that if $H^1_{\mathrm{dR}}(M) = 0$ (e.g. if $\pi_1(M)$ is finite) then any closed 1-form ω on M is exact: $\omega = df$. If ω is nowhere vanishing, the function f has no critical points. Hence the foliation given by ω is the foliation given by the submersion f. For example, the Reeb foliation on S^3, which is clearly not given by a submersion, is hence not given by a closed 1-form.

In general, the integrability condition $d\omega \wedge \omega = 0$ for a nowhere vanishing 1-form ω implies that locally $\omega = gdf$ for a submersion f which locally defines the foliation.

(4) Let $(M, \mathcal{F})$ and $(M', \mathcal{F}')$ be foliated manifolds. Then a (smooth) map $f \colon M \to M'$ preserves the foliation structure (hence it is a map of foliated manifolds) if and only if $df(T(\mathcal{F})) \subset T(\mathcal{F}')$.

Let $(M, \mathcal{F})$ be a foliated manifold and $T(\mathcal{F})$ the corresponding tangent bundle of $\mathcal{F}$. We say that $\mathcal{F}$ is *orientable* if the tangent bundle $T(\mathcal{F})$ is orientable, and that $\mathcal{F}$ is *transversely orientable* if its *normal bundle* $N(\mathcal{F}) = T(M)/T(\mathcal{F})$ is orientable. An *orientation* of $\mathcal{F}$ is an orientation of $T(\mathcal{F})$, and a *transverse orientation* of $\mathcal{F}$ is an orientation of $N(\mathcal{F})$.

Exercises 1.3 (1) Show that a foliation $\mathcal{F}$ is transversely orientable if and only if it can be represented by a Haefliger cocycle (U_i, s_i, γ_{ij}) with the property that

$$\det(d\gamma_{ij})_y > 0$$

for any $y \in s_j(U_i \cap U_j)$.

(2) Show that a foliation of codimension 1 is given by a nowhere vanishing integrable 1-form (or a nowhere vanishing vector field) if and only if it is transversely orientable.

(3) Determine which of the foliations in Examples 1.1 are orientable and which are transversely orientable.

(4) Find an example of a foliation of dimension 1 of the Klein bottle, which is neither orientable nor transversely orientable.

Let $\mathcal{F}$ be a transversely orientable foliation of codimension 1 on M. Hence $\mathcal{F}$ is given by an integrable nowhere vanishing differential 1-form ω on M. The form ω is determined uniquely up to the multiplication by a nowhere vanishing smooth function on M.

We have mentioned above that the condition $d\omega \wedge \omega = 0$ implies that $d\omega = \alpha \wedge \omega$. The form α is not uniquely determined, but we shall see that

(i) $d\alpha \wedge \omega = 0$ and $d(\alpha \wedge d\alpha) = 0$,

(ii) the class $\mathrm{gv}(\omega) = [\alpha \wedge d\alpha] \in H_{\mathrm{dR}}^3(M)$ is independent of the choice of α, and

(iii) $\mathrm{gv}(\omega) = \mathrm{gv}(h\omega)$ for any nowhere vanishing smooth function h on M.

It follows the class $\mathrm{gv}(\omega)$ depends only on the foliation $\mathcal{F}$ and not on the particular choice of ω or α. This class is called the *Godbillon–Vey class* of the foliation $\mathcal{F}$, and is denoted by

$$\mathrm{gv}(\mathcal{F}) \in H_{\mathrm{dR}}^3(M)\ .$$

Let us now prove the properties (i), (ii) and (iii).

(i) Since $d\omega = \alpha \wedge \omega$, we have

$$
\begin{aligned}
0 &= dd\omega \\
&= d(\alpha \wedge \omega) \\
&= d\alpha \wedge \omega - \alpha \wedge d\omega \\
&= d\alpha \wedge \omega - \alpha \wedge \alpha \wedge \omega \\
&= d\alpha \wedge \omega\ .
\end{aligned}
$$

As before, this implies $d\alpha = \gamma \wedge \omega$ for some 1-form γ. In particular,

$$d(\alpha \wedge d\alpha) = d\alpha \wedge d\alpha = \gamma \wedge \omega \wedge \gamma \wedge \omega = 0\ .$$

(ii) Let $\alpha' \in \Omega^1(M)$ be another form satisfying $d\omega = \alpha' \wedge \omega$. It follows that $(\alpha' - \alpha) \wedge \omega = 0$, so $\alpha' - \alpha = f\omega$ for a smooth function f on M. Hence

$$\alpha' \wedge d\alpha' = (\alpha + f\omega) \wedge d\alpha' = \alpha \wedge d\alpha' + f\omega \wedge d\alpha'\ .$$

Note that $\omega \wedge d\alpha' = 0$ by (i). Thus

$$\alpha' \wedge d\alpha' = \alpha \wedge d\alpha'$$

$$= \alpha \wedge d(\alpha + f\omega)$$
$$= \alpha \wedge d\alpha + \alpha \wedge d(f\omega)$$
$$= \alpha \wedge d\alpha - d(\alpha \wedge f\omega) .$$

The last equation follows from $d(\alpha \wedge f\omega) = d\alpha \wedge f\omega - \alpha \wedge d(f\omega)$ and part (i).

(iii) First we compute

$$
\begin{aligned}
d(h\omega) &= dh \wedge \omega + h d\omega \\
&= \frac{1}{h} dh \wedge h\omega + \alpha \wedge h\omega \\
&= (d(\log|h|) + \alpha) \wedge h\omega .
\end{aligned}
$$

So with $\alpha'' = d(\log|h|) + \alpha$ we have $\mathrm{gv}(h\omega) = [\alpha'' \wedge d\alpha'']$. But

$$\alpha'' \wedge d\alpha'' = (d(\log|h|) + \alpha) \wedge d\alpha = \alpha \wedge d\alpha + d(\log|h| + d\alpha) .$$

1.3 Constructions of foliations

In this section we list some standard constructions of foliations.

Product of foliations. Let $(M, \mathcal{F})$ and $(N, \mathcal{G})$ be two foliated manifolds. Then there is the *product foliation* $\mathcal{F} \times \mathcal{G}$ on $M \times N$, which can be constructed as follows. If $\mathcal{F}$ is represented by a Haefliger cocycle (U_i, s_i, γ_{ij}) on M and $\mathcal{G}$ is represented by a Haefliger cocycle $(V_k, s'_k, \gamma'_{kl})$ on N, then $\mathcal{F} \times \mathcal{G}$ is represented by the Haefliger cocycle

$$(U_i \times V_k, s_i \times s'_k, \gamma_{ij} \times \gamma'_{kl})$$

on $M \times N$. We have $\mathrm{codim}(\mathcal{F} \times \mathcal{G}) = \mathrm{codim}\,\mathcal{F} + \mathrm{codim}\,\mathcal{G}$ and $T(\mathcal{F} \times \mathcal{G}) = T(\mathcal{F}) \oplus T(\mathcal{G}) \subset T(M) \oplus T(N) = T(M \times N)$.

Pull-back of a foliation. Let $f \colon N \to M$ be a smooth map and $\mathcal{F}$ a foliation of M of codimension q. Assume that f is transversal to $\mathcal{F}$: this means that f is transversal to all the leaves of $\mathcal{F}$, i.e. for any $x \in N$ we have

$$(df)_x(T_x(N)) + T_{f(x)}(\mathcal{F}) = T_{f(x)}(M) .$$

Then we get a foliation $f^*(\mathcal{F})$ of N as follows.

Suppose that $\mathcal{F}$ is given by the Haefliger cocycle (U_i, s_i, γ_{ij}) on M. Put $V_i = f^{-1}(U_i)$ and $s'_i = s_i \circ f|_{V_i}$. The maps s'_i are submersions. To see this, take any $x \in V_i$. We have to show that

$$(ds'_i)_x = (ds_i)_{f(x)} \circ (df)_x$$

is surjective. But $(ds_i)_{f(x)}$ is surjective and trivial on $T_{f(x)}(\mathcal{F})$, hence it factors through the quotient $w\colon T_{f(x)}(M) \to T_{f(x)}(M)/T_{f(x)}(\mathcal{F})$ as a surjective map. Also $w \circ (df)_x$ is surjective since f is transversal to the leaves, and hence $(ds_i')_x$ is surjective as well. The foliation $f^*(\mathcal{F})$ is now given by the Haefliger cocycle (V_i, s_i', γ_{ij}) on N. We have codim $f^*(\mathcal{F}) = $ codim $\mathcal{F}$ and $T(f^*(\mathcal{F})) = df^{-1}(T(\mathcal{F}))$.

Transverse orientation cover of a foliation. For a foliated manifold $(M, \mathcal{F})$ put

$$\mathrm{toc}(M, \mathcal{F}) = \{(x, \mathcal{O}) \mid x \in M, \ \mathcal{O} \text{ orientation of } N_x(\mathcal{F})\} \ .$$

There is an obvious smooth structure on $\mathrm{toc}(M, \mathcal{F})$ such that the projection $p\colon \mathrm{toc}(M, \mathcal{F}) \to M$ is a twofold covering projection, called the *transverse orientation cover* of the foliated manifold $(M, \mathcal{F})$. The lift $\mathrm{toc}(\mathcal{F}) = p^*(\mathcal{F})$ of $\mathcal{F}$ to the transverse orientation cover is a transversely orientable foliation.

Orientation cover of a foliation. For any foliated manifold $(M, \mathcal{F})$ there is also a smooth structure on

$$\mathrm{oc}(M, \mathcal{F}) = \{(x, \mathcal{O}) \mid x \in M, \ \mathcal{O} \text{ orientation of } T_x(\mathcal{F})\}$$

such that the projection $p\colon \mathrm{toc}(M, \mathcal{F}) \to M$ is a twofold covering projection. This covering space is called the *orientation cover* of the foliated manifold $(M, \mathcal{F})$. The lift $\mathrm{oc}(\mathcal{F}) = p^*(\mathcal{F})$ of $\mathcal{F}$ to the orientation cover is an orientable foliation.

Exercises 1.4 (1) Show that if $\mathcal{F}$ is a foliation of an orientable manifold M then $\mathcal{F}$ is orientable if and only if $\mathcal{F}$ is transversely orientable.

(2) Find an example of non-orientable foliation of dimension 1 on the torus. What is the orientation cover of that foliation?

(3) By using the (transverse) orientation cover, show that if a compact manifold M carries a foliation of dimension 1 (or of codimension 1) then the Euler characteristic of M is 0. In particular, the only closed surfaces which admit a foliation of dimension 1 are the torus and the Klein bottle.

Quotient foliation. Let $(M, \mathcal{F})$ be a foliated manifold, and let G be a group acting freely and properly discontinuously by diffeomorphisms on M, so that the quotient manifold M/G is Hausdorff. We assume that the foliation $\mathcal{F}$ is *invariant* under this action of G, which means that any diffeomorphism $g\colon M \to M$ in G maps leaves to leaves, or equivalently, that $dg(T(\mathcal{F})) = T(\mathcal{F})$ for any $g \in G$. Then $\mathcal{F}$ induces a foliation $\mathcal{F}/G$ of M/G in the following way.

First denote by $p\colon M \to M/G$ the quotient map, which is a covering projection. Let $(\varphi_i\colon U_i \to \mathbb{R}^{n-q} \times \mathbb{R}^q)$ be a foliation atlas of $\mathcal{F}$. We may assume that $p|_{U_i}$ is injective for any i, by replacing (φ_i) by a refinement if necessary. Then

$$(\varphi_i \circ (p|_{U_i})^{-1}\colon p(U_i) \longrightarrow \mathbb{R}^{n-q} \times \mathbb{R}^q)$$

is a foliation atlas representing $\mathcal{F}/G$. If L is a leaf of $\mathcal{F}$, then the *isotropy group* $G_L = \{g \in G \,|\, g(L) = L\}$ of L acts smoothly on L, and the orbit manifold L/G_L can be identified with a leaf of $\mathcal{F}/G$ via the natural immersion of L/G_L into M/G. We have $\mathrm{codim}(\mathcal{F}/G) = \mathrm{codim}(\mathcal{F})$ and $T(\mathcal{F}/G) = dp(T(\mathcal{F}))$. Observe that we already used this construction in Example 1.1 (3).

Suspension of a diffeomorphism. This is another example of a quotient foliation. Let $f\colon F \to F$ be a diffeomorphism of a manifold F. The space $\mathbb{R} \times F$ has the obvious foliation of dimension 1, by the leaves $\mathbb{R} \times \{x\}$, $x \in F$. The smooth action of $\mathbb{Z}$, defined on $\mathbb{R} \times F$ by

$$(k, (t, x)) \mapsto (t + k, f^k(x)) ,$$

$k \in \mathbb{Z}$, $t \in \mathbb{R}$, $x \in F$, is properly discontinuous and it maps leaves to leaves. Thus we obtain the quotient foliation $\mathcal{S}_f$ on the (Hausdorff) manifold $(\mathbb{R} \times F)/\mathbb{Z} = \mathbb{R} \times_\mathbb{Z} F$. The foliated manifold $(\mathbb{R} \times_\mathbb{Z} F, \mathcal{S}_f)$ is called the *suspension* of the diffeomorphism f.

Foliation associated to a Lie group action. We first recall some terminology. For a smooth action $G \times M \to M$, $(g, x) \mapsto gx$, of a Lie group G on a smooth manifold M, the *isotropy* (or *stabilizer*) *subgroup* at $x \in M$ is the subgroup $G_x = \{g \in G \,|\, gx = x\}$. It is a closed subgroup of G, hence itself a Lie group. The *orbit* of x is $Gx = \{gx \,|\, g \in G\}$. It can be viewed as a manifold injectively immersed into M, via the immersion $G/G_x \to M$ with the image Gx.

We say that the action of G on M is *foliated* if $\dim(G_x)$ is a constant function of x. In this case the connected components of the orbits of the action are leaves of a foliation of M. As an integrable subbundle of $T(M)$, this foliation can simply be described in terms of the Lie algebra $\mathfrak{g}$ of G, namely as the image of the derivative of the action, which is a map of vector bundles $\mathfrak{g} \times M \to T(M)$ of constant rank.

In the case $G = \mathbb{R}$, a smooth $\mathbb{R}$-action on M is called a *flow* on M.

To such an action μ: $\mathbb{R} \times M \to M$ one can associate a vector field X on M by

$$X(x) = \frac{\partial \mu(t, x)}{\partial t}\bigg|_{t=0} .$$

A non-trivial flow μ is foliated precisely if its associated vector field X vanishes nowhere; in this case the foliation with the orbits of μ is the foliation given by the line field corresponding to X.

Exercise 1.5 Let $R \subset M \times M$ be an equivalence relation on a manifold M. By Godement's theorem (see Serre (1965)), M/R is a smooth manifold whenever R is a submanifold of $M \times M$ and $\mathrm{pr}_2 \colon R \to M$ is a submersion. Formulate and prove a result which gives sufficient conditions for a foliation $\mathcal{F}$ on M to induce a foliation on M/R.

Flat bundles. The following method of constructing foliations is related to the previous one of quotient foliations, and prepares the reader for the treatment of Reeb stability in Section 2.3.

Let p: $E \to M$ be a (smooth) fibre bundle over a connected manifold M. Then p is in particular a submersion, and thus defines the foliation $\mathcal{F}(p)$ of E whose leaves are the connected components of the fibres of p, i.e. the leaves are 'vertical'.

Sometimes it is also possible to construct a foliation of E with 'horizontal' leaves, so that p maps each leaf to M as a covering projection. The following construction captures these examples.

Let $G = \pi_1(M, x)$ be the fundamental group of M at a base-point $x \in M$, and let $\tilde{M}$ be the universal cover of M; or, more generally, suppose that G is any group acting freely and properly discontinuously on a connected manifold $\tilde{M}$ such that $\tilde{M}/G = M$. We will write the action of G on $\tilde{M}$ as a right action. Suppose also that there is a left action by G on a manifold F. Now form the quotient space

$$E = \tilde{M} \times_G F ,$$

obtained from the product space $\tilde{M} \times F$ by identifying (yg, z) with (y, gz) for any $y \in \tilde{M}$, $g \in G$ and $z \in F$. Thus E is the orbit space of $\tilde{M} \times F$ with respect to a properly discontinuous action of G. It is also Hausdorff, so it is a manifold. The projection pr_1: $\tilde{M} \times F \to \tilde{M}$ induces a submersion

$\pi\colon E \to M$, so we have the following commutative diagram:

$$
\begin{array}{ccc}
\tilde{M} \times F & \longrightarrow & E = \tilde{M} \times_G F \\
\downarrow{\scriptstyle \mathrm{pr.}} & & \downarrow{\scriptstyle \pi} \\
\tilde{M} & \longrightarrow & M
\end{array}
$$

The map $\pi\colon E \to M$ has the structure of a fibre bundle over M with fibre F.

Exercise 1.6 Show that the fibre bundles which can be obtained in this way are exactly the fibre bundles with discrete structure group.

The foliation $\mathcal{F}(\mathrm{pr}_2)$ of $\tilde{M} \times F$, which is given by the submersion $\mathrm{pr}_2\colon \tilde{M} \times F \to F$, is invariant under the action of G and hence we obtain the quotient foliation $\mathcal{F} = \mathcal{F}(\mathrm{pr}_2)/G$ on E. If $z \in F$ and $G_z \subset G$ is the isotropy group at z of the action by G on F, then the leaf of E obtained from the leaf $\tilde{M} \times \{z\}$ is naturally diffeomorphic to $\tilde{M}/G_z$, and π restricted to this leaf is the covering $\tilde{M}/G_z \to M$ of M.

The suspension of a diffeomorphism discussed above is a special case of this construction.

2

Holonomy and stability

In this chapter, we will present the stability theorems for foliations, which were proved by Ehresmann and Reeb when the theory of foliated manifolds was first developed in the 1950s. These theorems express that, under certain conditions, all the leaves near a given leaf look identical. Central to these theorems is the notion of holonomy. The basic idea of holonomy goes back to Poincaré, in his study of the 'first return' map of vector fields. For a given point x on a manifold equipped with a foliation of codimension q, one may consider how the leaves near that point intersect a small q-dimensional disk which is transversal to the leaves and contains the given point. The ways in which these leaves depart from this disk and return to it are encoded in a group, called the holonomy group at x. It is a quotient group of the fundamental group of the leaf through x. This group contains a lot of information about the structure of the foliation around the leaf through x, especially if that leaf is compact. For example, if this group is finite then all the nearby leaves must also be compact, and the foliation locally looks like one which is obtained by the flat bundle construction from the previous chapter. This is the 'local' Reeb stability theorem, discussed in Section 2.3 below. In Section 2.5, we will present a 'global' stability theorem, which applies to foliations of codimension 1, and states that under certain conditions, the holonomy group has to be trivial and the foliation has to be simple (in the technical sense of being given by the fibres of a submersion). In Section 2.6 we also give Thurston's more refined version of these stability theorems.

The notion of holonomy is closely related to that of a 'transverse' Riemannian structure on the foliation. Any manifold can be equipped with a Riemannian metric, but it is a special property for a foliated manifold to be equipped with a Riemannian metric for which the length of tangent vectors or curves which are transversal to the leaves is invariant under the holonomy group. In this case, the holonomy group can be interpreted as a group of isometries of a transversal q-dimensional disk as above. Riemannian metrics with this extra property were isolated and studied by Reinhart (1959), who called them 'bundle-like metrics'.

Foliated manifolds equipped with such a bundle-like metric are called 'Riemannian foliations'. Such foliations will be studied in more detail in Chapter 4, but we already introduce them here, because their properties are intimately related to those of the holonomy group. For example, any Riemannian foliation with compact leaves necessarily has finite holonomy groups and satisfies the conditions of the local Reeb stability theorem. Conversely, any foliation with finite holonomy groups and compact leaves can be equipped with the structure of a Riemannian foliation.

For such a foliation with finite holonomy and compact leaves, the local Reeb stability theorem implies that the space of leaves locally has the same structure as that of the space of orbits of a finite group acting on a Euclidean open ball. Spaces with such a local structure occur in many other contexts as well, and were studied by Satake under the name 'V-manifolds', and later by Thurston who called them 'orbifolds'. In Section 2.4 we make explicit the relation between orbifolds and foliations, and show among other things that, conversely, any orbifold can be obtained as the space of leaves of a foliation with compact leaves and finite holonomy.

2.1 Holonomy

The notion of holonomy uses that of a germ of a locally defined diffeomorphism. Let M and N be manifolds, and $x \in M$ and $y \in N$ any points. A *germ* of a map from x to y is an equivalence class of maps $f: U \to V$ from an open neighbourhood U of x to an open neighbourhood V of y with $y = f(x)$. Two such maps $f: U \to V$ and $f': U' \to V'$ determine the same germ from x to y if there exists an open neighbourhood $W \subset U \cap U'$ of x such that $f|_W = f'|_W$. We denote the germ of f from x to y (or the germ of f at x) by f_x or $\mathrm{germ}_x(f): (M, x) \to (N, y)$. Note that germs $f_x: (M, x) \to (N, y)$ and $g_y: (N, y) \to (O, z)$ can be composed into $g_y \circ f_x = (g \circ f|_{\mathrm{dom}g})_x: (M, x) \to (O, z)$, that there is an *identity germ* $\mathrm{id}_x: (M, x) \to (M, x)$. A germ of a (locally defined) diffeomorphism from x to y is a germ at x of a map $f: U \to V$ with $f(x) = y$ as above which is a diffeomorphism. Note that each germ of a diffeomorphism $f_x: (M, x) \to (N, y)$ has an inverse $f_x^{-1} = (f^{-1})_x: (N, y) \to (M, x)$. In particular, the germs of diffeomorphisms $(M, x) \to (M, x)$ form a group, denoted by

$$\mathrm{Diff}_x(M) \ .$$

We denote by $\mathrm{Diff}_x^+(M)$ the subgroup of $\mathrm{Diff}_x(M)$ of germs of diffeomorphisms which preserve orientation at x.

Now let $(M, \mathcal{F})$ be a foliated manifold, with $q = \mathrm{codim}(\mathcal{F})$, and let L be a leaf of $\mathcal{F}$. Let $x, y \in L$ be two points on this leaf, and let T and S be *transversal sections* at x and y (i.e. submanifolds of M transversal to the leaves of $\mathcal{F}$, with $x \in T$ and $y \in S$). To any path α from x to y in L we will associate a germ of a diffeomorphism

$$\mathrm{hol}(\alpha) = \mathrm{hol}^{S,T}(\alpha) \colon (T, x) \longrightarrow (S, y) \ ,$$

called the *holonomy* of the path α in L with respect to the transversal sections T and S, as follows.

First assume that there exists a (domain of a) foliation chart U of $\mathcal{F}$ such that $\alpha([0, 1]) \subset U$. In particular, the points x and y lie on the same plaque in U. Then we can find a small open neighbourhood A of x in T with $A \subset U$ on which we can define a smooth map $f \colon A \to S$ which satisfies the following: $f(x) = y$, and for any $x' \in A$ the point $f(x')$ lies on the same plaque in U as x'. Clearly we can choose A so small that f is a diffeomorphism onto its image. Then we define

$$\mathrm{hol}^{S,T}(\alpha) = \mathrm{germ}_x(f) \ .$$

Clearly this definition does not depend on the choice of U and f. Observe that if β is another path in $L \cap U$ from x to y then $\mathrm{hol}^{S,T}(\alpha) = \mathrm{hol}^{S,T}(\beta)$.

Now we shall consider the general case. Choose a sequence of foliation charts $U_1, \ldots, U_k$ so that $\alpha([\frac{i-1}{k}, \frac{i}{k}]) \subset U_i$. Let α_i be a path in $L \cap U_i$ from $\alpha(\frac{i-1}{k})$ to $\alpha(\frac{i}{k})$, and choose transversal sections T_i of $\mathcal{F}$ at $\alpha(\frac{i}{k})$, with $T_0 = T$ and $T_k = S$ (Figure 2.1). Now define

$$\mathrm{hol}^{S,T}(\alpha) = \mathrm{hol}^{T_k, T_{k-1}}(\alpha_k) \circ \cdots \circ \mathrm{hol}^{T_1, T_0}(\alpha_1) \ .$$

Again it is easy to see that this definition is independent of the choice of the sequence $U_1, \ldots, U_k$ of foliation charts, so it really depends only on α and the transversal sections T and S. The germ $\mathrm{hol}^{S,T}(\alpha)$ may again be represented by a diffeomorphism $f \colon A \subset T \to S$ with the property that $f(x')$ lies on the same leaf as an $x' \in A$. Sometimes we will denote such a diffeomorphism f by $\mathrm{hol}^{S,T}(\alpha)$ as well.

Let us list some basic properties of holonomy:

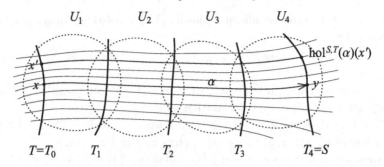

Fig. 2.1. Holonomy $(k = 4)$

(i) If α is a path in L from x to y and β is another path in L from y to z, and if T, S and R are transversal sections of $\mathcal{F}$ at x, y and z respectively, then

$$\text{hol}^{R,T}(\beta\alpha) = \text{hol}^{R,S}(\beta) \circ \text{hol}^{S,T}(\alpha) \ .$$

Here $\beta\alpha$ denotes the concatenation of α and β.

(ii) If α and β are homotopic paths in L (with fixed end-points) from x to y and if T and S are transversal sections at x respectively y, then

$$\text{hol}^{S,T}(\alpha) = \text{hol}^{S,T}(\beta) \ .$$

Hence we can regard $\text{hol}^{S,T}$ as being defined on the homotopy classes of paths in L from x to y. (To see this, first observe that if β is very close to α one can use the same sequence of foliation charts $U_1, \ldots, U_k$ to define both $\text{hol}^{S,T}(\alpha)$ and $\text{hol}^{S,T}(\beta)$, which clearly implies that these two germs are equal. The general case follows by decomposing the homotopy between α and β into a sequence of 'small' ones.)

(iii) Let α be a path in L from x to y, let T and T' be two transversal sections at x and let S and S' be two transversal sections at y. Then

$$\text{hol}^{S',T'}(\alpha) = \text{hol}^{S',S}(\bar{y}) \circ \text{hol}^{S,T}(\alpha) \circ \text{hol}^{T,T'}(\bar{x}) \ .$$

Here $\bar{x}$ denotes the constant path with image x. In this sense, the holonomy is independent of the choice of the transversal sections.

From these basic properties, we see in particular that for a transversal section T at $x \in L$ one obtains the map

$$\mathrm{hol}^T = \mathrm{hol}^{T,T} \colon \pi_1(L, x) \longrightarrow \mathrm{Diff}_x(T)$$

which is a group homomorphism. Since $\mathrm{Diff}_x(T) \cong \mathrm{Diff}_0(\mathbb{R}^q)$ and because of (iii) above, we obtain a homomorphism of groups

$$\mathrm{hol} \colon \pi_1(L, x) \longrightarrow \mathrm{Diff}_0(\mathbb{R}^q)$$

which is called the *holonomy homomorphism* of L, and is determined uniquely up to a conjugation in $\mathrm{Diff}_0(\mathbb{R}^q)$. In particular, the image $\mathrm{Hol}(L, x)$ of hol, which is called the *holonomy group* of L, is determined up to an inner automorphism of $\mathrm{Diff}_0(\mathbb{R}^q)$. Thus we have the exact sequence

$$1 \longrightarrow K \longrightarrow \pi_1(L, x) \xrightarrow{\ \mathrm{hol}\ } \mathrm{Hol}(L, x) \longrightarrow 1 \,,$$

where K is the kernel of hol.

One says that two paths α and β from x to y in L *have the same holonomy* if $\mathrm{hol}(\alpha\beta^{-1}) = 1$. This is an equivalence relation, which is well-defined also on the homotopy classes of paths in L from x to y. The equivalence class of α with respect to this relation is also referred to as the *holonomy class* of α.

By taking the differential at 0 of a (germ of a) diffeomorphism, one obtains a homomorphism of groups $d_0 \colon \mathrm{Diff}_0(\mathbb{R}^q) \to GL(q, \mathbb{R})$. The composition

$$d\mathrm{hol} = d_0 \circ \mathrm{hol} \colon \pi_1(L, x) \longrightarrow GL(q, \mathbb{R})$$

is called the *linear holonomy homomorphism* of L, while its image is called the *linear holonomy group* of L and is denoted by $d\mathrm{Hol}(L, x)$.

Exercises 2.1 (1) Check that $\mathrm{Hol}(L, x)$ (and hence also $d\mathrm{Hol}(L, x)$) is independent of the choice of the base-point x up to a conjugation in $\mathrm{Diff}_0(\mathbb{R}^q)$. (For this reason, one often omits the base-point from the notation.)

(2) Show that if $\mathcal{F}$ is transversely orientable then the holonomy group $\mathrm{Hol}(L, x)$ of a leaf of $\mathcal{F}$ is a subgroup of $\mathrm{Diff}_0^+(\mathbb{R}^q)$.

Examples 2.2 You should try to see yourself what the group $\mathrm{Hol}(L, x)$ looks like in any example you meet.

(1) If the leaf L is simply connected, the holonomy group of L is trivial.

(2) For a leaf L of the foliation $\mathcal{F}(f)$ given by a submersion $f\colon M \to N$ the holonomy group of L is trivial.

(3) For the foliation of the Möbius band (Example 1.1 (4)), whose leaves are circles, each leaf has trivial holonomy, except the central one whose holonomy group is $\mathbb{Z}_2 = \mathbb{Z}/2\mathbb{Z}$.

(4) Consider the Reeb foliation on the solid torus. All the leaves are simply connected except for the boundary leaf T^2. It is easy to see that the definition of holonomy still makes sense for a boundary leaf, except that a transversal section at a point of the leaf has the boundary as well. For example, in our case we can choose that the transversal section is diffeomorphic to (and identified with) the interval $[0, \infty)$. Now if α and β are the generators of the fundamental group of the boundary torus, one has $\mathrm{hol}(\alpha) = 1$, while $\mathrm{hol}(\beta)$ is a germ from 0 to 0 of a diffeomorphism $f\colon [0, \infty) \to [0, \infty)$ with $f(t) < t$ for any $t > 0$.

The Reeb foliation on S^3 has one compact leaf T^2, and the holonomy group of this leaf is $\mathbb{Z} \oplus \mathbb{Z}$. For α and β as above $\mathrm{hol}(\alpha)(t)$ and $\mathrm{hol}(\beta)(t)$ are germs from 0 to 0 of diffeomorphisms $g, h\colon \mathbb{R} \to \mathbb{R}$ such that

$$
\begin{aligned}
g(t) &< t \quad \text{for} \quad t < 0\,, \\
g(t) &= t \quad \text{for} \quad t \geq 0\,,
\end{aligned}
$$

and

$$
\begin{aligned}
h(t) &= t \quad \text{for} \quad t \leq 0\,, \\
h(t) &< t \quad \text{for} \quad t > 0\,.
\end{aligned}
$$

Exercises 2.3 (1) If $\pi_1(M)$ acts on F we get a foliation on $\tilde{M} \times_{\pi_{\cdot}(M)} F = E$; see Section 1.3. The space E is a fibre bundle over M with fibre F which is a transversal section of the foliation:

$$
\begin{array}{ccc}
\tilde{M} \times F & \xrightarrow{\ p\ } & E = \tilde{M} \times_{\pi_{\cdot}(M)} F \\
\downarrow & & \downarrow \\
\tilde{M} & \longrightarrow & M
\end{array}
$$

If L is a leaf through $p(y, z)$ then $L \cong \tilde{M}/\pi_1(M)_z$ where $\pi_1(M)_z$ is the isotropy at z of the action of $\pi_1(M)$ on F. Show that $\mathrm{Hol}(L, p(y, z))$ is the image of the restriction of the action of $\pi_1(M)$ to $\pi_1(M)_z$, as a map $\pi_1(M)_z \to \mathrm{Diff}_z(F)$.

(2) Let G act on M and let $\mathcal{F}$ be a foliation of M invariant under the action. Let L be a leaf of $\mathcal{F}$ and let L' be the corresponding leaf of the quotient foliation $\mathcal{F}/G$ (see Section 1.3). Express the holonomy of L' in terms of the holonomy of L and of the action of G.

(3) Let $(M, \mathcal{F})$ be a foliated manifold of codimension q, L a leaf of $\mathcal{F}$, $\alpha \colon [0,1] \to L$ a path, and let T and S be transversal sections at $\alpha(0)$ and $\alpha(1)$ respectively, and fix an embedding $Z_0 \colon \mathbb{R}^q \to T$ for which $Z_0(0) = \alpha(0)$.

Now consider the mappings

$$Z \colon [0,1] \times B \longrightarrow M$$

with the following properties:

(i) B is an open disk in $\mathbb{R}^q$ centred at 0,

(ii) $Z(0, \text{-}) = Z_0|_B$ and $Z(1, B) \subset S$,

(iii) $Z(t, \text{-})$ is a smooth embedding transversal to $\mathcal{F}$, for any $t \in [0,1]$,

(iv) $\alpha_y = Z(\text{-}, y)$ is a path in a leaf L_y, for any $y \in B$, and $\alpha_0 = \alpha$.

Show that such a map Z exists. Observe that $\mathrm{hol}^{S,T}(\alpha)(Z(0,y)) = Z(1,y)$. Show that if $Z' \colon [0,1] \times B' \to M$ is another such map, then $Z(\text{-}, y)$ and $Z'(\text{-}, y)$ are in the same homotopy class in L_y with fixed end-points, for any y close to 0.

2.2 Riemannian foliations

The aim of this section is to present a very brief introduction to Riemannian foliations.

First let us introduce some notations. Let M be a manifold of dimension n. A symmetric $C^\infty(M)$-bilinear form

$$g \colon \mathfrak{X}(M) \times \mathfrak{X}(M) \longrightarrow C^\infty(M)$$

is said to be *positive* if it satisfies $g(X, X) \geq 0$ for any $X \in \mathfrak{X}(M)$. Such a form induces a positive bilinear form g_x on the tangent space $T_x(M)$, at any point $x \in M$. The kernel $\mathrm{Ker}(g_x)$ is the linear subspace $\{\xi \in T_x(M) \mid g_x(\xi, T_x(M)) = 0\}$ of $T_x(M)$. The *Lie derivative* $L_X g$ of g in the direction of a vector field $X \in \mathfrak{X}(M)$ is the symmetric $C^\infty(M)$-bilinear form on $\mathfrak{X}(M)$ given by

$$L_X g(Y, Z) = X(g(Y, Z)) - g([X, Y], Z) - g(Y, [X, Z]) \, .$$

Let $\mathcal{F}$ be a foliation of codimension q of the manifold M. A *transverse metric* on $(M, \mathcal{F})$ is a positive $C^\infty(M)$-bilinear form g on $\mathfrak{X}(M)$ such that

(i) $\mathrm{Ker}(g_x) = T_x(\mathcal{F})$ for any $x \in M$, and

(ii) $L_X g = 0$ for any vector field X on M tangent to $\mathcal{F}$.

A foliation $\mathcal{F}$ of M together with a transverse metric g on $(M, \mathcal{F})$ is called a *Riemannian foliation* of M.

Remark 2.4 Note that the condition (i) means exactly that g is the pull-back of a Riemannian structure on the normal bundle $N(\mathcal{F})$ along the standard projection $T(M) \to N(\mathcal{F})$. Furthermore we note that (i) alone implies that $T_x(\mathcal{F}) \subset \mathrm{Ker}((L_X g)_x)$ for any vector field X tangent to $\mathcal{F}$ (i.e. any section of $T(\mathcal{F})$) and any $x \in M$.

To understand the condition (ii), which is clearly local, consider a surjective foliation chart $\varphi = (x_1, \ldots, x_{n-q}, y_1, \ldots, y_q) \colon U \to \mathbb{R}^{n-q} \times \mathbb{R}^q$ of $\mathcal{F}$. In this chart, the form g is determined by the functions

$$g_{ij} = g\left(\frac{\partial}{\partial y_i}, \frac{\partial}{\partial y_j} \right).$$

Now if X is a vector field on M tangent to $\mathcal{F}$, then $[X, \frac{\partial}{\partial y_i}]$ is again a section of $T(\mathcal{F})|_U$ and therefore

$$L_X g\left(\frac{\partial}{\partial y_i}, \frac{\partial}{\partial y_j} \right) = X(g_{ij}).$$

Hence $L_X g|_U = 0$ for any such X if and only if

$$\frac{\partial g_{ij}}{\partial x_k} = 0$$

for $i, j = 1, \ldots, q$ and $k = 1, \ldots, n-q$. In other words, the functions g_{ij} are constant along the plaques in U, i.e. they are functions of the coordinates $y_1, \ldots, y_q$. Equivalently, $g|_U$ is the pull-back of a Riemannian metric on $\mathbb{R}^q$ along the submersion $\mathrm{pr}_2 \circ \varphi \colon U \to \mathbb{R}^q$.

Let $(\mathcal{F}, g)$ be a Riemannian foliation of M. If T is a transversal section of $(M, \mathcal{F})$, the restriction $g|_T$ is a Riemannian metric on T. So any transversal section of a Riemannian foliation has a natural Riemannian structure.

For a given foliation $\mathcal{F}$ on M, a Riemannian structure on the normal bundle of $\mathcal{F}$ determines a transverse metric if and only if this structure is holonomy invariant. One half of this is stated in the following proposition, the other half in Remark 2.7 (2).

Proposition 2.5 *Let $(\mathcal{F}, g)$ be a Riemannian foliation of M. Let L be a leaf of $\mathcal{F}$, α a path in L, and let T and S be transversal sections of $\mathcal{F}$ with $\alpha(0) \in T$ and $\alpha(1) \in S$. Then*

$$\mathrm{hol}^{S,T}(\alpha) \colon (T, \alpha(0)) \longrightarrow (S, \alpha(1))$$

is the germ of an isometry.

Proof We have to prove that $h = \mathrm{hol}^{S,T}(\alpha)$ preserves the metric. By the definition of holonomy, we can assume that α is inside a surjective foliation chart $\varphi = (x_1, \ldots, x_{n-q}, y_1, \ldots, y_q) \colon U \to \mathbb{R}^{n-q} \times \mathbb{R}^q$ of $\mathcal{F}$ and that $T, S \subset U$. We can also assume without loss of generality that $\varphi(T) \subset \{0\} \times \mathbb{R}^q$, so that the vector fields $\frac{\partial}{\partial y_i}|_T$ form a frame for the tangent bundle of T. Furthermore let us assume that the holonomy diffeomorphism $h \colon T \to S$ is defined on all of T. By the definition of h we have

$$y_i \circ h = y_i|_T$$

for $i = 1, \ldots, q$. Therefore $\frac{\partial(y_i \circ h)}{\partial y_j}(p) = \delta_{ij}$ for $i, j = 1, \ldots, q$, so

$$dh_p\left(\frac{\partial}{\partial y_i}(p)\right) \in \frac{\partial}{\partial y_i}(h(p)) + T_{h(p)}(\mathcal{F})$$

for any $p \in T$. Here we view $T_{h(p)}(S)$ as a subspace of $T_{h(p)}(M)$. In particular,

$$
\begin{aligned}
g|_S\left(dh_p\left(\frac{\partial}{\partial y_i}(p)\right), dh_p\left(\frac{\partial}{\partial y_j}(p)\right)\right) &= g\left(\frac{\partial}{\partial y_i}(h(p)), \frac{\partial}{\partial y_j}(h(p))\right) \\
&= g_{ij}(h(p)) \\
&= g_{ij}(p) \\
&= g|_T\left(\frac{\partial}{\partial y_i}(p), \frac{\partial}{\partial y_j}(p)\right).
\end{aligned}
$$

$\square$

Theorem 2.6 *Let $(\mathcal{F}, g)$ be a Riemannian foliation of M and assume that all the leaves of $\mathcal{F}$ are compact. Then each leaf has a finite holonomy group.*

Proof Let L be a leaf of $\mathcal{F}$ and $x \in L$, and let S be a transversal section of $(M, \mathcal{F})$ with $x \in S$. Since S has its natural Riemannian structure, we have the exponential map

$$\exp_x \colon B(0, \varepsilon) \longrightarrow S,$$

which is an open embedding defined on the open ε-ball centred at 0 with respect to the inner product $(g|_S)_x$ on $T_x(S)$, and satisfies $\exp_x(0) = x$. Denote the image of $\exp_x$ by U. We may assume that U is relatively compact in S. Represent any element of the holonomy group $H = \mathrm{Hol}(L, x)$

by a holonomy embedding $h\colon (V_h, x) \to (U, x)$, for an open neighbourhood $V_h \subset U$ of x. By Proposition 2.5 we know that h is an isometry. This implies that we may choose V_h so that $h(V_h) = V_h = \exp_x(B(0, \delta_h))$ for some small positive $\delta_h \leq \varepsilon$, and furthermore it implies that $d_x h$ is an orthogonal transformation of $T_x(S)$ satisfying $\exp_x \circ d_x h = h \circ \exp_x$. Since L is compact, the group H is finitely generated, and therefore we may assume that all the neighbourhoods V_h are equal to say $V = \exp_x(B(0, \delta))$. Thus we have represented H as a group of isometries of (V, x).

Consider an orbit of H in V. Since H acts on V by holonomy diffeomorphisms, the orbit lies in a leaf L of $\mathcal{F}$. The intersection $L \cap S$ is discrete, by compactness of L. Since U, and hence V, is relatively compact in S, it follows that $L \cap V$ is finite. This implies that the orbit is finite as well.

By taking the differential at x, or equivalently, by conjugating with $\exp_x$, we faithfully represent the group H as a group of orthogonal transformations of $T_x(S)$, with the property that any orbit of the action of H in $T_x(S)$ is finite. It is now an easy exercise in linear algebra to see that such a group has to be finite.   □

Remarks 2.7 (1) Let $\mathcal{F}$ be a simple foliation of M given by a surjective submersion $M \to N$ with connected fibres. By taking the pull-back along this submersion one obtains a bijective correspondence between the Riemannian metrics on N and the transverse metrics on $(M, \mathcal{F})$.

(2) Let $\mathcal{F}$ be a foliation of M given by a Haefliger cocycle (U_i, s_i, γ_{ij}). If each submersion $s_i\colon U_i \to s_i(U_i)$ has connected fibres, then any transverse metric on $(M, \mathcal{F})$ induces a Riemannian metric on $s_i(U_i)$, for any i, such that the diffeomorphisms γ_{ij} are isometries. Conversely, if each $s_i(U_i)$ is a Riemannian manifold and if each γ_{ij} is an isometry, then the pull-back of the Riemannian structure on $s_i(U_i)$ along s_i gives a transverse metric on $(U_i, \mathcal{F}|_{U_i})$, and these transverse metrics amalgamate to a transverse metric on $(M, \mathcal{F})$.

(3) If $(\mathcal{F}, g)$ is a Riemannian foliation of M and $(\mathcal{F}', g')$ a Riemannian foliation of N, then the product of g and g' gives a transverse metric on $(M \times M', \mathcal{F} \times \mathcal{F}')$. Thus the product of Riemannian foliations is again a Riemannian foliation.

(4) The pull-back of a Riemannian foliation is again a Riemannian foliation in the natural way. This is obvious if one uses the description of a Riemannian foliation by a Haefliger cocycle of isometries, as in (2).

(5) Let $\mathcal{F}/G$ be the quotient foliation of the manifold M/G obtained

as the orbit space of a free and properly discontinuous action of a group G on M by diffeomorphisms which preserve the foliation. If g is a transverse metric on $(M, \mathcal{F})$ such that the diffeomorphisms of the action preserve g, then g induces a transverse metric on $(M/G, \mathcal{F}/G)$.

(6) Let M be a manifold, and let G be a group acting freely and properly discontinuously on a manifold $\tilde{M}$ such that $\tilde{M}/G = M$. Let F be a Riemannian manifold equipped with a (left) action of G by isometries. We obtain the flat bundle $E = \tilde{M} \times_G F$ with its natural foliation $\mathcal{F}$, as in Section 1.3. The pull-back of the Riemannian metric on F along the projection $\mathrm{pr}_2 \colon \tilde{M} \times F \to F$ gives a transverse metric on $(\tilde{M} \times F, \mathcal{F}(\mathrm{pr}_2))$, by (1). This metric is clearly preserved by the action of G on $\tilde{M} \times F$, so it induces a transverse metric on the quotient foliation $(E, \mathcal{F})$ by (5). Thus the flat bundle foliation is in this case Riemannian. In particular, the suspension of an isometry $f \colon F \to F$ is a Riemannian foliation.

(7) Let $\mathcal{F}$ be a foliation of M, and let $\langle\,\text{-}\,,\,\text{-}\,\rangle$ be a Riemannian metric on M. Consider the canonical decomposition of the tangent bundle of M,

$$T(M) = T(\mathcal{F}) \oplus T(\mathcal{F})^\perp \,,$$

and note that $T(\mathcal{F})^\perp$ is isomorphic to $N(\mathcal{F})$. According to this decomposition let us write a vector field X on M as the sum of its tangent and normal components, $X = X^{(t)} + X^{(n)}$. We say that the vector field X is *normal* to $\mathcal{F}$ if $X = X^{(n)}$. By decomposing the metric in the same way we obtain a positive $C^\infty(M)$-bilinear form g on M with $\mathrm{Ker}(g_x) = T_x(\mathcal{F})$, that is

$$g(X, Y) = \langle X^{(n)}, Y^{(n)} \rangle \,.$$

The metric $\langle\,\text{-}\,,\,\text{-}\,\rangle$ is called *bundle-like* with respect to $\mathcal{F}$ if the associated form g is a transverse metric on $(M, \mathcal{F})$.

One can express this in terms of projectable vector fields. A vector field Y on M is called *projectable* with respect to $\mathcal{F}$ if $[X, Y]$ is tangent to $\mathcal{F}$, for any vector field X tangent to $\mathcal{F}$. Intuitively, this means that the normal component of Y is constant along each leaf. In the literature, the projectable vector fields are also referred to as *foliated*, or *basic* vector fields. Global projectable vector fields which are not tangent to $\mathcal{F}$ may be rare, but locally there are plenty of them (e.g. the fields $\frac{\partial}{\partial y_i}$ in Remark 2.4 are projectable). Note that the normal component of a projectable vector field is projectable. It is straightforward to see that a Riemannian metric $\langle\,\text{-}\,,\,\text{-}\,\rangle$ on M is bundle-like with respect to $\mathcal{F}$ if and only if for any vector field X tangent to $\mathcal{F}$ and any two normal projectable vector fields Y and Z defined on an open subset U of M we have $X(\langle Y, Z \rangle) = 0$.

(8) Let M be a manifold equipped with a foliated action of a Lie group G, and let $\mathcal{F}$ be the associated foliation of M. Now assume $\langle\,$-$\,,\,$-$\,\rangle$ is a Riemannian metric on M such that G acts on M by isometries. Then $\langle\,$-$\,,\,$-$\,\rangle$ is bundle-like with respect to $\mathcal{F}$, i.e. the associated form g is a transverse metric on $(M, \mathcal{F})$, so $(\mathcal{F}, g)$ is a Riemannian foliation of M.

In fact, one can prove that $L_X g = 0$ for any vector field X tangent to $\mathcal{F}$ (this is clearly sufficient). If $G = \mathbb{R}$ and if the action is non-trivial, then G acts as the one parameter group of isometries of the associated non-singular vector field X tangent to $\mathcal{F}$. Thus g is preserved by this group of isometries, or equivalently, $L_X g = 0$. In general, one can find k one-dimensional subgroups of G (by using the exponential map in G) such that the associated vector fields $X_1, \ldots, X_k$ on M locally, over some open subset U of M, form a frame of $T(\mathcal{F})|_U$. Then $L_X. g|_U = \cdots = L_{X_k} g|_U = 0$ as above.

Proposition 2.8 *Let $\mathcal{F}$ be the foliation of a manifold M given by a foliated action of a compact Lie group G. Then there exists a bundle-like Riemannian metric on M with respect to $\mathcal{F}$, and any leaf of $\mathcal{F}$ is compact with finite holonomy group.*

Proof Take any Riemannian metric $\langle\,$-$\,,\,$-$\,\rangle$ on M, and define a new one by taking the average over G with respect to the Haar measure μ on G,

$$\rho(X, Y) = \int_G \langle g_* X, g_* Y \rangle \, d\mu(g) \,, \qquad X, Y \in \mathfrak{X}(M) \,.$$

Here $g_* X$ stands for the composition of X with the derivative of the action of $g \in G$ on M. Now G acts on M by isometries with respect to ρ. By Remark 2.7 (8) it follows that ρ is bundle-like with respect to $\mathcal{F}$. Any leaf of $\mathcal{F}$ is compact since G is, and has finite holonomy by Theorem 2.6. $\qquad\square$

2.3 Local Reeb stability

In this section, $\mathcal{F}$ is a fixed foliation of codimension q on a manifold M of dimension n, and L denotes a fixed leaf of $\mathcal{F}$. Write H for the holonomy group $H = \mathrm{Hol}(L, x_0)$, where $x_0 \in L$, so that there is an exact sequence

$$1 \longrightarrow K \longrightarrow \pi_1(L, x_0) \xrightarrow{\mathrm{hol}} H \longrightarrow 1 \,.$$

Let $\tilde{L} \to L$ be the covering space of L corresponding to the subgroup K (the *holonomy cover* of L). So H acts freely on $\tilde{L}$ and $\tilde{L}/H \cong L$.

Assume now that L is compact and H is finite. Observe that $\tilde{L}$ is compact also in this case. Let T be a transversal section of $\mathcal{F}$ at x_0, so small that each element of H can be represented by a holonomy diffeomorphism of T (which maps a point y of T to a point which lies on the same leaf as y, see definition in Section 2.1). This can be done since H is finite: Represent each element $h \in H$ by a holonomy embedding $f_h \colon U_h \to T'$, where U_h is an open neighbourhood of x_0 in an arbitrary transversal section T' at x_0. Then choose an open neighbourhood W of x_0 in T' so small that $W \subset U_h$ and $f_h \circ f_k|_W = f_{hk}|_W$, for all $h, k \in H$. Finally, take $T = \bigcap_{h \in H} f_h(W)$. Note that T can be chosen arbitrarily small.

Theorem 2.9 (Local Reeb stability) *For a compact leaf L with finite holonomy as above, there exist a saturated open neighbourhood V of L in M and a diffeomorphism*

$$\tilde{L} \times_H T \longrightarrow V$$

under which the foliation $\mathcal{F}$ restricted to V corresponds to the flat bundle foliation on $\tilde{L} \times_H T$.

REMARK. A subset of M is called *saturated* if it is a union of leaves of $\mathcal{F}$. The flat bundle foliation is discussed in Section 1.3. Observe that $\tilde{L} \times_H T \cong L' \times_{\pi_.(L,x_.)} T$ where L' is the universal cover of L and $\pi_1(L, x_0)$ acts on T via the holonomy homomorphism $\pi_1(L, x_0) \to H$.

By the differentiable slice theorem, actions by finite groups can be linearized locally (in the smooth context). So we can indeed assume that $T \cong \mathbb{R}^q$ and that H acts linearly on T; we will not use this here.

NOTATION. Points of $\tilde{L}$ can be denoted as holonomy equivalence classes $[\alpha]$ of paths α from x_0 to x in L: two paths α and β from x_0 to x in L are in the same holonomy class if the homotopy class of $\beta^{-1}\alpha$ is in K. Then the action by H on $\tilde{L}$ can be denoted as a right action, by

$$[\alpha][\eta] = [\alpha\eta] \,,$$

where η is a loop at x_0 in L representing an element $[\eta] = \mathrm{hol}(\eta) \in H$.

Proof (of Theorem 2.9) Let us first fix a retraction r, a finite cover $\mathcal{U}$, a number c and a transversal section T in the following convenient way.

(i) Take a tubular neighbourhood N of L in M, with corresponding retraction $r\colon N \to L$, so small that $r^{-1}(x)$ is transversal to the leaves of $\mathcal{F}$, for any $x \in L$ (this can always be achieved by shrinking an arbitrary tubular neighbourhood of L). Then $T_x = r^{-1}(x)$ defines a transversal section at x, so we have chosen transversal sections T_x 'smoothly in x'.

(ii) Choose a finite cover $\mathcal{U}$ of L by domains of foliation charts of $\mathcal{F}$. We assume that for any $U \in \mathcal{U}$ we have $U \subset N$, while the corresponding foliation chart ($\varphi\colon U \to \mathbb{R}^{n-q} \times \mathbb{R}^q$) has image of the form $B \times \mathbb{R}^q$ for a simply connected open subset B of $\mathbb{R}^{n-q}$ and satisfies $\varphi(r(x)) = (\mathrm{pr}_1(\varphi(x)), 0)$ for any $x \in U$. In particular, each $U \in \mathcal{U}$ intersects L in only one simply connected plaque. We can further assume that the intersection $U \cap U' \cap L$ for any two elements U and U' of $\mathcal{U}$ is either connected or empty. Also pick a specific element $U_0 \in \mathcal{U}$ with $x_0 \in U_0$. Now the homotopy (and hence the holonomy) classes of paths in L from x_0 to x can be represented by *chains* in $\mathcal{U}$ from x_0 to x, i.e. by sequences $(U_0, U_1, \ldots, U_k)$ of elements of $\mathcal{U}$ satisfying $x \in U_k$ and $U_{i-1} \cap U_i \neq \emptyset$ for any $i = 1, 2, \ldots, k$. Fix a number c so that

(a) every $x \in L$ can be reached by a chain $(U_0, U_1, \ldots, U_k)$ in $\mathcal{U}$ from x_0 to x of length $k + 1 \leq c$ (such a c exists by compactness of L), and

(b) every $[\beta] \in H$ can be represented by a chain $(U_0, U_1, \ldots, U_k = U_0)$ in $\mathcal{U}$ of length $k + 1 \leq c$ (such a c exists by finiteness of H).

Observe now that the following property is a consequence:

(∗) every point $p \in \tilde{L}$ can be represented by a chain $(U_0, U_1, \ldots, U_k)$ in $\mathcal{U}$ and a point $x \in U_k \cap L$, where $k + 1 \leq 2c$.

Indeed, to see this, let $p \in \tilde{L}$ and write $p = [\alpha]$ for a path α from x_0 to x in L. By property (a) there exists a chain $(U_0, U_1, \ldots, U_k)$ in $\mathcal{U}$ of length $k + 1 \leq c$ with $x \in U_k$. This chain represents the homotopy class of a path β in L from x_0 to x. Then $\beta^{-1}\alpha$ is a loop at x_0, and the holonomy class of $\beta^{-1}\alpha$ can be represented by a chain of length $\leq c$ (by property (b)). Therefore $[\alpha] = [\beta(\beta^{-1}\alpha)]$ can be represented by a chain of length $\leq 2c$.

(iii) Now take a transversal section $T \subset T_x$. at x_0 so small that H acts as a group of holonomy diffeomorphisms on T (as explained above), and so small that for every chain ζ in $\mathcal{U}$ from x_0 to x of length $\leq 6c$, the 'holonomy transport'

$$\mathrm{hol}^{T_x, T}(\zeta)\colon T \longrightarrow T_x$$

is defined as an embedding of all of T into T_x. This holonomy map was defined explicitly in Section 2.1 using the foliation charts corresponding to the sets in the chain ζ, and the properties of these foliation charts listed in (ii) guarantee that such a section T exists. Moreover, we can take T to be so small that for any two chains ζ and ζ' from x_0 to x of length $\leq 6c$ which represent the same holonomy class we have $\mathrm{hol}^{T_x,T}(\zeta) = \mathrm{hol}^{T_x,T}(\zeta')$. In addition, we take T to be so small that for any chain ζ' from x_0 to x_0 of length $\leq 4c$ the map $\mathrm{hol}^{T_x,\,T}(\zeta)$ is exactly the chosen action on T of the holonomy class represented by the chain ζ'. And finally, we take T so small that for any chain ζ from x_0 to x of length $\leq 2c$ and any chain ζ' from x_0 to x_0 of length $\leq 4c$ we have $\mathrm{hol}^{T_x,T}(\zeta\zeta') = \mathrm{hol}^{T_x,T}(\zeta)\mathrm{hol}^{T_x,\,T}(\zeta')$.

With all this fixed, we can now define a 'transport'

$$\tilde{\sigma}\colon \tilde{L} \times T \longrightarrow M$$

by

$$\tilde{\sigma}([\alpha], y) = \mathrm{hol}^{T_x,T}(\zeta)(y) \in T_x \ ,$$

where ζ is any chain in $\mathcal{U}$ of length $\leq 2c$ representing $[\alpha]$ and $x = \alpha(1)$. Such a ζ exists by $(*)$, and the definition is independent of the choice of ζ by property (iii). Let us now observe the following.

(1) From the local form of $\mathcal{F}$, it is obvious that $\tilde{\sigma}$ is a local diffeomorphism (it is a diffeomorphism on plaques in a given chain). In particular, $\tilde{\sigma}(\,\cdot\,, y)\colon \tilde{L} \to L_y$ is an open map to the leaf L_y through y. It is also closed since $\tilde{L}$ is compact, hence onto. In fact it is a covering projection (exercise: any proper local diffeomorphism is a covering projection). Thus the image of $\tilde{\sigma}$ is a saturated open subset V of M,

$$V = \bigcup_{y \in T} L_y \ .$$

(2) The map $\tilde{\sigma}$ factors through the quotient as $\sigma\colon \tilde{L} \times_H T \to V \subset M$:

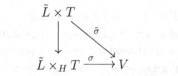

Indeed, if $[\alpha]$ is represented by a chain ζ from x_0 to x and $h \in H$ is represented by a chain ζ' from x_0 to x_0, both of length $\leq 2c$, then (iii) implies

$$\tilde{\sigma}([\alpha]h, y) \quad = \quad \mathrm{hol}^{T_x,T}(\zeta\zeta')(y)$$

$$= \operatorname{hol}^{T_x,T}(\zeta)(\operatorname{hol}^{T_x,\,\cdot\,,T}(\zeta')(y))$$
$$= \operatorname{hol}^{T_x,T}(\zeta)(hy)$$
$$= \tilde{\sigma}([\alpha], hy) \ .$$

The map σ is a local diffeomorphism, since the map $\tilde{L} \times T \to \tilde{L} \times_H T$ is a covering projection.

(3) The map σ is injective. To this end, suppose $y, y' \in T$ and α and α' are two paths representing points of $\tilde{L}$ with end-points x and x' respectively, so that

$$\tilde{\sigma}([\alpha], y) = \tilde{\sigma}([\alpha'], y') \ .$$

Then for the retraction $r \colon N \to L$ of the tubular neighbourhood we have

$$x = r(\tilde{\sigma}([\alpha], y)) = r(\tilde{\sigma}([\alpha'], y')) = x' \ ,$$

so α and α' have the same end-point $x = x'$. Next, represent $[\alpha]$ and $[\alpha']$ by chains ζ and ζ' respectively, both of length $\leq 2c$, and let $\gamma = \alpha^{-1}\alpha'$. Note that $[\gamma]$ is represented by the chain $\zeta^{\mathrm{op}}\zeta'$, i.e. the concatenation of ζ' and of the order-reversed ζ. Property (iii) implies

$$\begin{aligned}
\operatorname{hol}^{T_x,T}(\zeta)([\gamma]y') &= \operatorname{hol}^{T_x,T}(\zeta)(\operatorname{hol}^{T_x,\,\cdot\,,T}(\zeta^{\mathrm{op}}\zeta')(y')) \\
&= \operatorname{hol}^{T_x,T}(\zeta\zeta^{\mathrm{op}}\zeta')(y') \\
&= \operatorname{hol}^{T_x,T}(\zeta')(y') \\
&= \tilde{\sigma}([\alpha'], y') \\
&= \tilde{\sigma}([\alpha], y) \\
&= \operatorname{hol}^{T_x,T}(\zeta)(y) \ .
\end{aligned}$$

Thus $[\gamma]y' = y$, so $([\alpha'], y') = ([\alpha][\gamma], y')$ and $([\alpha], y) = ([\alpha], [\gamma]y')$ are in the same equivalence class in $\tilde{L} \times_H T$. $\qquad\square$

2.4 Orbifolds

In this section we will see that as a consequence of the local Reeb stability theorem, the space of leaves of a foliation with compact leaves with finite holonomy has a natural orbifold structure (Theorem 2.15). In particular this is true for the space of orbits of a foliated compact Lie group action. And conversely, any orbifold is isomorphic to the orbifold given by such an action (Propositions 2.22 and 2.23).

The space of leaves $M/\mathcal{F}$ of a foliated manifold often contains little or no information about the foliation itself. For example, the space of leaves of the Kronecker foliation of the torus has the trivial topology.

What we know in general is that the quotient projection $M \to M/\mathcal{F}$ is open. On the other hand, for a simple foliation given by a surjective submersion $M \to N$ with connected fibres the space of leaves is just the manifold N and reflects the 'transverse' structure of the foliation entirely. Further, for the standard foliation of the flat bundle $\tilde{M} \times_G F$ described in Section 1.3 the space of leaves is the orbit space F/G. According to the local Reeb stability theorem we can conclude that locally around a compact leaf with finite holonomy the space of leaves is the orbit space of a smooth action of a finite group.

Let $(M, \mathcal{F})$ be a foliated manifold for which all the leaves are compact with finite holonomy. Then the preceding observation holds for any leaf, so there is a (countable) cover (V_i) of M by saturated open sets such that V_i is diffeomorphic to the flat bundle $\tilde{L}_i \times_{H_i} T_i$ by a diffeomorphism which maps $\mathcal{F}|_{V_i}$ to the standard foliation of the flat bundle. Here $\tilde{L}_i$ is the holonomy cover of the (compact) leaf L_i with (finite) holonomy group H_i and T_i a suitable transversal section; we may assume that T_i is diffeomorphic to an open subset U_i of $\mathbb{R}^q$. Now for any i we have the open map $\phi_i \colon T_i \to M/\mathcal{F}$, which is the restriction of the quotient map $M \to M/\mathcal{F}$ and induces an open embedding $T_i/H_i \to M/\mathcal{F}$. The sets $\phi_i(T_i) \cong T_i/H_i$ form an open cover of $M/\mathcal{F}$. So $M/\mathcal{F}$ is locally the orbit space of a smooth action of a finite group. It is easy to see that $M/\mathcal{F}$ is locally compact, Hausdorff and second-countable. Let i and j be such that $V_i \cap V_j \neq \emptyset$, and choose $x \in T_i$ and $y \in T_j$ with $\phi_i(x) = \phi_j(y)$. In particular x and y lie on the same leaf L, so there is a path in L from x to y. This path induces the holonomy diffeomorphism $h \colon W \to W'$ from an open neighbourhood W of x in T_i to an open neighbourhood W' of y in T_j. We have $\phi_j \circ h = \phi_i|_W$ since h preserves the leaves. As we shall see, this property gives us a compatibility between the local representations of $\phi_i(T_i)$ as the orbit spaces T_i/H_i.

NOTATION. Denote by $\mathrm{Diff}(M)$ the group of diffeomorphisms of a manifold M. Let G be a subgroup of $\mathrm{Diff}(M)$. Recall that the isotropy group of $x \in M$ is $G_x = \{g \in G \mid gx = x\}$. For any $g \in G$ put $\Sigma_g = \{x \in M \mid gx = x\}$, and write

$$\Sigma_G = \{x \in M \mid G_x \neq 1\} = \bigcup_{\mathrm{id} \neq g \in G} \Sigma_g \,.$$

A subset S of M is called *G-stable* if it is connected and if for any $g \in G$ we have either $gS = S$ or $gS \cap S = \emptyset$. The *isotropy group* of S is $G_S = \{g \in G \mid gS = S\}$. Observe that G-stable subsets of M are exactly the components of G-invariant subsets of M. If G is finite, the open G-stable subsets of M form a base for the topology of M. In fact, for any $x \in M$ we can find an arbitrarily small open G-stable neighbourhood S of x such that $G_x = G_S$.

First we shall recall some facts about the finite subgroups of $\mathrm{Diff}(M)$, for a manifold M. Take a finite subgroup G of $\mathrm{Diff}(M)$, and let $x \in M$. We can choose a G-invariant Riemannian metric on M, by Proposition 2.8. The exponential map associated to the metric gives us a diffeomorphism from an open ε-ball centred at 0 in the tangent space $T_x(M)$ to an open neighbourhood W of x, $\exp_x \colon B(0, \varepsilon) \to W \subset M$. Since the metric is G-invariant, $(dg)_x$ is an orthogonal transformation of $T_x(M)$ and $\exp_x \circ (dg)_x = g \circ \exp_x$, for any $g \in G_x$. In particular, if $(dg)_x = \mathrm{id}$ then $g|_W = \mathrm{id}$.

Lemma 2.10 *Let M be a connected manifold and G a finite subgroup of $\mathrm{Diff}(M)$. Then Σ_G is closed with empty interior and the differential $d_x \colon G_x \to \mathrm{Aut}(T_x M)$ is injective for each $x \in M$.*

REMARK. Here d_x is given by $d_x g = (dg)_x$ for any $g \in G_x$. Note that the lemma implies that any diffeomorphism of finite order on a connected manifold which fixes an open set is the identity.

Proof (of Lemma 2.10) Take $g \in G_x$ with $(dg)_x = \mathrm{id}$. Observe that $Z = \{y \in U \mid gy = y, (dg)_y = \mathrm{id}\}$ is a non-empty closed subset of M which is also open by the argument above the lemma. This proves the last statement of the lemma, which in particular implies that Σ_g has an empty interior, for any $g \in G - \{\mathrm{id}\}$. Then the same is true for the set Σ_G. $\qquad\square$

Lemma 2.11 *Let M be a manifold and G a finite subgroup of $\mathrm{Diff}(M)$. For any smooth map $f \colon V \to M$ defined on a non-empty open connected subset V of M, satisfying $f(x) \in Gx$ for each $x \in V$, there exists a unique $g \in G$ such that $f = g|_V$.*

Proof Observe that $A = M - \Sigma_G$ is an open dense G-invariant subset of M (by Lemma 2.10) and that the quotient projection $\pi \colon A \to A/G$ is a principal G-bundle. Hence $A \cap V$ is open and dense in V. Let C be a component of $A \cap V$. By assumption $f|_C \colon C \to A$ satisfies $\pi \circ f|_C = \pi|_C$, thus there is a unique $g_C \in G$ such that $f|_C = g_C|_C$. In particular, $(df)_x = (dg_C)_x$ for any $x \in \bar{C}$.

Note that the argument above the lemma shows that in a suitable chart around a point $y \in \Sigma_G \cap V$ the action of G_y is linear, and hence in this chart Σ_G is just a finite union of linear subspaces. In particular, there are finitely many components of $A \cap V$ intersecting this chart, and y lies in the boundary of all of them. But if C and C' are any two components of $A \cap V$ for which $y \in \bar{C} \cap \bar{C}'$, we have $(dg_C)_y = (df)_y = (dg_{C'})_y$, and hence $g_C = g_{C'}$ by Lemma 2.10. Therefore f coincides with a single $g \in G$ on a neighbourhood of y. Since V is connected, it follows that f equals g on all of V. $\qquad\square$

Let Q be a topological space. An *orbifold chart* of dimension $n \geq 0$ on Q is a triple (U, G, ϕ), where U is a connected open subset of $\mathbb{R}^n$, G is a finite subgroup of $\mathrm{Diff}(U)$ and $\phi\colon U \to Q$ is an open map which induces a homeomorphism $U/G \to \phi(U)$.

If (U, G, ϕ) is an orbifold chart on Q and S an open G-stable subset of U, the triple $(S, G_S, \phi|_S)$ is again an orbifold chart called the *restriction* of (U, G, ϕ) on S. More generally, let (V, H, ψ) be another orbifold chart on Q. An *embedding* $\lambda\colon (V, H, \psi) \to (U, G, \phi)$ between orbifold charts is an embedding $\lambda\colon V \to U$ such that $\phi \circ \lambda = \psi$. Let us list some basic properties of embeddings between orbifold charts:

Proposition 2.12 *(i) For any embedding* $\lambda\colon (V, H, \psi) \to (U, G, \phi)$ *between orbifold charts on Q, the image $\lambda(V)$ is a G-stable open subset of U, and there is a unique isomorphism $\bar{\lambda}\colon H \to G_{\lambda(V)}$ for which $\lambda(hx) = \bar{\lambda}(h)\lambda(x)$.*

(ii) The composition of two embeddings between orbifold charts is an embedding between orbifold charts.

(iii) For any orbifold chart (U, G, ϕ), any diffeomorphism $g \in G$ is an embedding of (U, G, ϕ) into itself, and $\bar{g}(g') = gg'g^{-1}$.

(iv) If $\lambda, \mu\colon (V, H, \psi) \to (U, G, \phi)$ are two embeddings between the same orbifold charts, there exists a unique $g \in G$ with $\lambda = g \circ \mu$.

Proof To prove (i), observe that for any $h \in H$ the diffeomorphism $\lambda \circ h \circ \lambda^{-1}$ of $\lambda(V)$ preserves the fibres of ϕ. By Lemma 2.11 there exists a unique $\bar{\lambda}(h) \in G$ which extends this diffeomorphism to U. The map $\bar{\lambda}\colon H \to G$ is a homomorphism by Lemma 2.10, and clearly injective.

Now take any $g \in G$. If $g \in \bar{\lambda}(H)$ we have $gV = V$. On the other hand, assume that $gV \cap V$ is non-empty. Then $A \cap gV \cap V$ is non-empty as well, where $A = U - \Sigma_G$, so there exists an $x \in A \cap V$ with $g^{-1}x \in V$. Since $\lambda^{-1}(x)$ and $\lambda^{-1}(g^{-1}x)$ are on the same fibre of ψ, there exists an $h \in H$ with $h\lambda^{-1}(g^{-1}x) = \lambda^{-1}(x)$ and hence $\bar{\lambda}(h)g^{-1}x = x$. Since $x \in A$, this yields $\bar{\lambda}(h) = g$, so $g \in \bar{\lambda}(H)$.

Now (ii) and (iii) are obvious, while (iv) follows from Lemma 2.11. □

We say that two orbifold charts (U, G, ϕ) and (V, H, ψ) of dimension n on Q are *compatible* if for any $z \in \phi(U) \cap \psi(V)$ there exist an orbifold chart (W, K, θ) on Q with $z \in \theta(W)$ and embeddings between orbifold charts $\lambda \colon (W, K, \theta) \to (U, G, \phi)$ and $\mu \colon (W, K, \theta) \to (V, H, \psi)$. There is an equivalent description of compatibility:

Proposition 2.13 *If two orbifold charts (U, G, ϕ) and (V, H, ψ) on Q are compatible, then any smooth map $f \colon Z \to V$ defined on an open subset $Z \subset U$ which satisfies $\psi \circ f = \phi|_Z$ is a local diffeomorphism. If in addition $Z = U$ then f is a covering projection onto its image, and its covering transformations form a subgroup of G. Conversely, if (U, G, ϕ) and (V, H, ψ) are two orbifold charts on Q and if for any $z \in \phi(U) \cap \psi(V)$ there exists an open subset $Z \subset U$ with $z \in \phi(U)$ and a smooth embedding $f \colon Z \to V$ with $\psi \circ f = \phi|_Z$, then (U, G, ϕ) and (V, H, ψ) are compatible.*

Proof Take any $x \in Z$. Since the charts are compatible, there exist a chart (W, K, θ) on Q with $\phi(x) \in \theta(W)$ and embeddings between orbifold charts $\lambda \colon (W, K, \theta) \to (U, G, \phi)$ and $\mu \colon (W, K, \theta) \to (V, H, \psi)$. By Proposition 2.12 (i) we can assume that W is in fact a G-stable subset of U and λ the inclusion, and we can also assume that $x \in W \subset Z$. Now $f \circ \mu^{-1}$ is an embedding by Lemma 2.11. This proves that f is a local diffeomorphism. If $Z = U$, the facts that ϕ is proper onto its image and V Hausdorff imply that f is a proper local diffeomorphism onto its image, and hence a covering projection. Any covering transformation of f is in G by Lemma 2.11.

For the second part of the statement observe that the restriction of f to a G-stable open set is an embedding between orbifold charts. □

An *orbifold atlas* of dimension n of a topological space Q is a collection of pairwise compatible orbifold charts

$$\mathcal{U} = \{(U_i, G_i, \phi_i)\}_{i \in I}$$

of dimension n on Q such that $\bigcup_{i \in I} \phi_i(U_i) = Q$. Two orbifold atlases of Q are *equivalent* if their union is an orbifold atlas.

An *orbifold* of dimension n is a pair $(Q, \mathcal{U})$, where Q is a second-countable Hausdorff topological space and $\mathcal{U}$ is a maximal orbifold atlas of dimension n of Q.

REMARK. Thus an orbifold is a space which is locally the orbit space of a smooth action of a finite group, but these actions are part of the structure. Observe that any orbifold atlas of a second-countable Hausdorff space Q is contained in a unique maximal orbifold atlas of Q, and hence defines an orbifold structure on Q. An orbifold $(Q, \mathcal{U})$ will often be denoted simply by Q, while the orbifold charts in this maximal atlas $\mathcal{U}$ will be referred to as orbifold charts of the orbifold Q. Also, any orbifold subatlas of $\mathcal{U}$ will be referred to as an orbifold atlas of the orbifold Q. It is clear that any orbifold is locally compact. Since it is also second-countable and Hausdorff, it follows that it is paracompact.

Note that any manifold chart on a manifold M may be viewed as an orbifold chart, with trivial group. Therefore a manifold is an example of an orbifold.

Exercise 2.14 Let Q be an orbifold of dimension n. Show that there exists an orbifold atlas $\mathcal{U}$ of Q such that $U = \mathbb{R}^n$ and G is a finite subgroup of $O(n)$, for any orbifold chart $(U, G, \phi) \in \mathcal{U}$.

Let Q and Q' be two orbifolds. A continuous map $f \colon Q \to Q'$ is an *orbifold map* if for any $z \in Q$ there exist orbifold charts (U, G, ϕ) of Q with $z \in \phi(U)$ and (V, H, ψ) of Q' and a smooth map $\tilde{f} \colon U \to V$ such that $\psi \circ \tilde{f} = f \circ \phi$ (such a map $\tilde{f}$ is called a *local lift* of f). Orbifold maps are closed under composition and form a category. Two orbifolds are *isomorphic* if they are isomorphic in this category. (There are other, more subtle notions of map between orbifolds, but they give rise to the same isomorphisms.) The orbifold maps $Q \to \mathbb{R}$ are called the *smooth functions* on the orbifold Q. Equivalently, a function $f \colon Q \to \mathbb{R}$ is smooth if $f \circ \phi$ is smooth for any chart (U, G, ϕ) in an orbifold atlas of Q.

Let Q be an orbifold, (U, G, ϕ) an orbifold chart of Q and $x \in U$, and write $z = \phi(x)$. By Lemma 2.10 the differential at x gives a faithful representation of G_x in $GL(n, \mathbb{R})$, and we denote the corresponding finite subgroup of $GL(n, \mathbb{R})$ by dG_x. Since $G_{gx} = gG_xg^{-1}$ for any $g \in G$, the points in the same orbit of G have isotropy groups in the same conjugacy class. In particular dG_x and dG_{gx} are in the same conjugacy class of $GL(n, \mathbb{R})$. Now let $\lambda \colon (V, H, \psi) \to (U, G, \phi)$ be an embedding between orbifold charts and $y \in V$ with $\lambda(y) = x$. Observe that $\bar{\lambda}(H_y) = G_x$ and that

$$dG_x = (d\lambda)_y dH_y (d\lambda)_y^{-1} .$$

Thus the conjugacy class of dG_x depends only on the point z and not on the choice of the orbifold chart (U, G, ϕ) on Q or on the choice of x. Therefore we can define the *isotropy group* of z, denoted by

$$\text{Iso}_z(Q) ,$$

as a subgroup of $GL(n, \mathbb{R})$ determined up to a conjugation.

Define the *singular locus* of Q by

$$\Sigma_Q = \{ z \in Q \mid \text{Iso}_z(Q) \neq 1 \} .$$

We have $\Sigma_Q \cap \phi(U) = \phi(\Sigma_G)$, and in particular Σ_Q is a closed subset of Q with empty interior.

From our discussion at the start of this section we can conclude:

Theorem 2.15 *Let $\mathcal{F}$ be a foliation of codimension q of a manifold M such that any leaf of $\mathcal{F}$ is compact with finite holonomy group. Then the space of leaves $M/\mathcal{F}$ has a canonical orbifold structure of dimension q. The isotropy group of a leaf in $M/\mathcal{F}$ is its holonomy group.*

Corollary 2.16 *Let M be a manifold equipped with a foliated action of a compact connected Lie group G. Then the orbit space M/G has a canonical orbifold structure.*

Proof By Proposition 2.8 all the leaves of the associated foliation $\mathcal{F}$ of M are compact with finite holonomy. Now Theorem 2.15 gives us the canonical orbifold structure on $M/\mathcal{F}$, which is M/G since G is connected. □

Exercise 2.17 Let G be a discrete group acting properly on a manifold M. Show that M/G has a canonical orbifold structure.

Exercise 2.18 Let G be a finite group acting on an orbifold Q by orbifold automorphisms. Show that Q/G has a natural orbifold structure. Extend this to the case where G is any discrete group acting properly on Q. Conclude that Corollary 2.16 also holds if G is not connected.

In the rest of this section we shall prove the following converse of Corollary 2.16.

Theorem 2.19 *Any orbifold is isomorphic to the orbifold associated to an action of a compact connected Lie group G with finite isotropy groups.*

In fact, we will show that we can choose G to be either $SO(n)$ in the orientable case (Proposition 2.22) or $U(n)$ in general (Proposition 2.23).

Let Q be an orbifold and let $\mathcal{U} = \{(U_i, G_i, \phi_i)\}_{i \in I}$ be the maximal orbifold atlas of Q. A Riemannian metric on the orbifold Q is a collection $\rho = (\rho_i)$, where ρ_i is a Riemannian metric on U_i, such that any embedding between orbifold charts $\lambda \colon (U_i, G_i, \phi_i) \to (U_j, G_j, \phi_j)$ is an isometry as a map $(U_i, \rho_i) \to (U_j, \rho_j)$.

REMARK. Equivalently one can define a Riemannian metric ρ to be given with respect to an arbitrary orbifold atlas of Q, but then the isometry condition should state that any embedding between orbifold charts $\lambda \colon (S, (G_i)_S, \phi_i|_S) \to (U_j, G_j, \phi_j)$ is an isometry, where S is any open G_i-stable subset of U_i. Then ρ can be uniquely extended to any other orbifold chart of Q such that the isometry condition still holds.

Proposition 2.20 *Any orbifold admits a Riemannian metric.*

Proof Let $\{(U_i, G_i, \phi_i)\}_{i \in I}$ be an orbifold atlas of Q. Since Q is paracompact, we may assume that the cover $(\phi_i(U_i))$ of Q is locally finite, and we may choose a partition of unity $(\alpha_i \colon \phi_i(U_i) \to \mathbb{R})$ subordinate to this cover. Following the standard proof of the existence of such a partition of unity, we can choose this partition to be smooth; explicitly, the map $\alpha_i \circ \phi_i$ is smooth for any i. Choose an arbitrary Riemannian metric $\langle \, \text{-} \, , \, \text{-} \, \rangle^{(i)}$ on U_i, for any i. Now for any i define a new Riemannian metric ρ_i on U_i as follows: for any $x \in U_i$ and any $\xi, \zeta \in T_x(U_i)$ put

$$(\rho_i)_x(\xi, \zeta) = \sum_{j \in I} \alpha_j(\phi_i(x)) \sum_{g \in G_j} \langle d(g \circ \lambda_j)_x(\xi), d(g \circ \lambda_j)_x(\zeta) \rangle^{(j)}_{g\lambda_j(x)} \, .$$

Here $\lambda_j \colon (S, (G_i)_S, \phi_i|_S) \to (U_j, G_j, \phi_j)$ is any embedding between orbifold charts defined on an open G_i-stable neighbourhood $S \subset U_i$ of x (the definition does not depend on the choice of this embedding because of the average by G_j and Proposition 2.12 (iv)). It is then straightforward to check that the collection (ρ_i) is a Riemannian metric on Q. $\square$

Let Q be an orbifold with the maximal orbifold atlas

$$\mathcal{U} = \{(U_i, G_i, \phi_i)\}_{i \in I},$$

and choose a Riemannian metric $\rho = (\rho_i)$ on Q. First we shall construct the *frame bundle* of Q as follows.

Recall that for a manifold M of dimension n, the *frame bundle* $F(M)$ is a smooth fibre bundle over M for which the fibre $F_x(M)$ is the manifold of all ordered bases of the tangent space $T_x(M)$. Any such basis may be viewed as the image of the standard basis under a unique isomorphism $e\colon \mathbb{R}^n \to T_x(M)$, so we may identify points of $F_x(M)$ with such isomorphisms. The frame bundle admits a canonical right action of the Lie group $GL(n, \mathbb{R})$ which makes it into a principal $GL(n, \mathbb{R})$-bundle over M. Explicitly the action is given by the composition, i.e. $eA = e \circ A$ for $e \in F_x(M)$ and $A \in GL(n, \mathbb{R})$.

First take an orbifold chart (U_i, G_i, ϕ_i) of Q, and consider the frame bundle $F(U_i)$. Note that this bundle is in fact trivial. Next note that the action of G_i lifts to a left action on $F(U_i)$, which is given by composition with the derivative dg, i.e. $ge = (dg)_x \circ e$ for any $e \in F_x(U_i)$. This action commutes with the right action of $GL(n, \mathbb{R})$ and is free; this follows immediately from Lemma 2.11. In particular it follows that $F(U_i)/G_i$ is a manifold equipped with a right action of $GL(n, \mathbb{R})$. Note that the map ϕ_i induces a map $p_i\colon F(U_i)/G_i \to Q$ in the obvious way and that the action of $GL(n, \mathbb{R})$ is along the fibres of this map.

Now take any embedding $\lambda\colon (U_i, G_i, \phi_i) \to (U_j, G_j, \phi_j)$ between orbifold charts. The composition with the derivative $d\lambda$ induces an embedding $\tilde{\lambda}\colon F(U_i) \to F(U_j)$. This embedding factors as

$$\lambda_*\colon F(U_i)/G_i \longrightarrow F(U_j)/G_j.$$

Indeed, for any $g \in G_i$ and $e \in T_x(U_i)$ we have

$$
\begin{aligned}
\tilde{\lambda}(ge) &= (d\lambda)_{gx} \circ (dg)_x \circ e \\
&= d(\lambda \circ g \circ \lambda^{-1})_{\lambda(x)} \circ (d\lambda)_x \circ e \\
&= d(\bar{\lambda}(g))_{\lambda(x)} \circ \tilde{\lambda}(e) \\
&= \bar{\lambda}(g)\tilde{\lambda}(e).
\end{aligned}
$$

Since $\lambda(U_i)$ is G_j-stable, the map λ_* is a smooth open embedding. In addition this embedding commutes with the action of $GL(n, \mathbb{R})$ and $p_j \circ \lambda_* = p_i$. Note that in particular $g_* = \mathrm{id}$, which implies that for any two embeddings $\lambda, \mu\colon (U_i, G_i, \phi_i) \to (U_j, G_j, \phi_j)$ we have $\lambda_* = \mu_*$.

We may conclude that the manifolds $F(U_i)/G_i$ (for all $i \in I$) together with the smooth open embeddings λ_* (for all embeddings between orbifold charts λ) form a filtered direct system. We define the *frame bundle* $F(Q)$ of the orbifold Q as the colimit of this system,

$$F(Q) = \varinjlim\{F(U_i)/G_i, \lambda_*\} \ .$$

Note that $F(Q)$ is a manifold, that each $F(U_i)/G_i$ is canonically embedded into $F(Q)$ as an open submanifold and that the maps p_i induce an open map $p\colon F(Q) \to Q$. Furthermore the Lie group $GL(n, \mathbb{R})$ acts smoothly on $F(Q)$ and this action is transitive along the fibres of p. In particular we have $F(Q)/GL(n, \mathbb{R}) \cong Q$. One can check that the isotropy groups of this action are all finite. In fact, for any $\eta \in F(Q)$ there is an isomorphism between $GL(n, \mathbb{R})_\eta$ and $\mathrm{Iso}_{p(\eta)}(Q)$. In particular the action is foliated, and the leaves of the associated foliation are of course the components of the fibres of p.

We may consider only the orthonormal frames on U_i with respect to the metric ρ_i, which give the *orthogonal frame bundle* $OF(U_i)$ on U_i. It is a principal $O(n)$-bundle over U_i. Since all the embeddings between orbifold charts are isometries, the same construction as before gives us the *orthogonal frame bundle* $OF(Q)$ of the orbifold Q. It comes with the right action of the Lie group $O(n)$ which is transitive along the fibres of p and has finite isotropy groups.

An *orientation* of Q is a decomposition of $F(Q)$ into two disjoint open submanifolds $F(Q) = F^+(Q) \cup F^-(Q)$ such that $p(F^+(Q)) = p(F^-(Q)) = Q$. We say that Q is *orientable* if such a decomposition exists. If Q is oriented, the fibres of $p\colon F^+(Q) \to Q$ are connected and $F^+(Q)$ is invariant under the action of $GL^+(n, \mathbb{R})$, while any matrix $A \in GL^-(n, \mathbb{R})$ maps $F^+(Q)$ onto $F^-(Q)$. The bundle $F^+(Q)$ is called the *positive frame bundle* of the orbifold Q.

Exercise 2.21 Equivalently, an orientation of an orbifold Q is given by an atlas $\{(V_k, H_k, \psi_k)\}$ of Q such that any embedding between orbifold charts $\lambda\colon (S, (H_k)_S, \psi_k|_S) \to (V_l, H_l, \psi_l)$ is orientation preserving, for each H_k-stable subset S of V_k.

Assume that Q is oriented. With respect to ρ we may also consider the *positive orthogonal frame bundle* $OF^+(Q) = OF(Q) \cap F^+(Q)$, which comes with the action of the connected compact Lie group $SO(n)$ transitive along the fibres of p and with finite isotropy groups.

We observe:

Proposition 2.22 *Let Q be an oriented orbifold of dimension n equipped with a Riemannian metric. Then Q is isomorphic to the orbifold associated to the foliated action of the group $SO(n)$ on the positive orthogonal frame bundle $OF^+(Q)$.*

Here the only non-trivial point is that the holonomy of a leaf is equal to the corresponding isotropy group, which is true because the Lie group is connected. This is why we need the orientability. For a non-orientable orbifold Q we may achieve the same thing by using the compact connected unitary group $U(n)$ instead. This is done by replacing the tangent bundle $T(U_i)$ by its complexification $T(U_i) \otimes \mathbb{C}$. This is now a complex vector bundle of rank n over U_i, and we may again consider the ordered (complex) bases in the fibres $T_x(U_i) \otimes \mathbb{C}$. Such a basis is now represented by a complex linear isomorphism $e \colon \mathbb{C}^n \to T_x(U_i) \otimes \mathbb{C}$. The collection of such bases yield the complex frame bundle $\mathbb{C}F(U_i)$ which is now a principal $GL(n, \mathbb{C})$-bundle. Again there is a free G_i-action on $\mathbb{C}F(U_i)$, now given by the composition with $dg \otimes \mathrm{id}_{\mathbb{C}}$, for $g \in G_i$. Similarly an embedding λ between orbifold charts induces an equivariant embedding $\tilde{\lambda}$ by the composition with $d\lambda \otimes \mathrm{id}_{\mathbb{C}}$. We can then repeat the same construction as before to obtain the *complex frame bundle* $\mathbb{C}F(Q)$ of the orbifold Q. This bundle is now equipped with a smooth action of the connected Lie group $GL(n, \mathbb{C})$.

Next, the metric ρ_i also extends naturally to a complex Riemannian structure on the bundle $T(U_i) \otimes \mathbb{C}$. So we may again consider the orthonormal frames only, and we get the *unitary frame bundle* $UF(Q)$ of the orbifold Q. It comes with the action of the compact connected group $U(n)$ which is transitive along the fibres of p and has finite isotropy groups. Now we have

Proposition 2.23 *Let Q be an orbifold of dimension n equipped with a Riemannian metric. Then Q is isomorphic to the orbifold associated to the foliated action of the group $U(n)$ on the unitary frame bundle $UF(Q)$.*

These two propositions in particular provide a proof of Theorem 2.19.

2.5 Global Reeb stability in codimension 1

In this section M is a compact connected manifold, equipped with a transversely orientable foliation $\mathcal{F}$ of codimension 1. We will prove

Theorem 2.24 (Global Reeb stability) *If $\mathcal{F}$ is a codimension 1 transversely orientable foliation of a compact connected manifold M and admits a compact leaf L_0 with finite fundamental group, then $\mathcal{F}$ is the foliation given by the fibres of a fibre bundle projection $\phi \colon M \to S^1$. In particular, all the leaves of $\mathcal{F}$ are diffeomorphic to L_0.*

The proof will be divided into various small steps. First, we show that the assumptions imply that the holonomy of L_0 is trivial.

Lemma 2.25 *If L is a leaf of a codimension 1 transversely orientable foliation $\mathcal{F}$ and if the fundamental group of L is finite, then the holonomy group of L is trivial.*

Proof Take a transversal section T at a point $x \in L$, and represent an arbitrary element g of the holonomy group of L by a smooth embedding $f \colon U \to T$ of an open neighbourhood $U \subset T$ of x, with $f(x) = x$. Since g has a finite order k, we can find an open neighbourhood $W \subset U$ of x such that $f(W) = W$ and $(f|_W)^k = \mathrm{id}$. Then $(df)_x$ is of finite order and preserves orientation, hence equals id. By Lemma 2.10 it follows that $f|_W = \mathrm{id}$. □

Now fix a finite atlas of $(M, \mathcal{F})$, consisting of surjective foliation charts $(\psi_i \colon W_i \to \mathbb{R}^{n-1} \times \mathbb{R})_{i=1}^{k}$ for which the sets $U_i = \psi_i^{-1}((-1,1)^n)$ cover M. Therefore the restrictions

$$(\varphi_i = \psi_i|_{U_i} \colon U_i \longrightarrow \mathbb{R}^{n-1} \times \mathbb{R})_{i=1}^{k}$$

form a finite foliation atlas of $(M, \mathcal{F})$.

The atlas (φ_i) has the following useful property: a leaf L of $\mathcal{F}$ is compact if and only if it meets each of the charts U_i in finitely many plaques. Indeed, if L hits U_i in infinitely many plaques, then the plaques of L in W_i together with $L - \psi_i^{-1}([-2,2]^n)$ would form an open cover of L without a finite subcover, thus L is not compact. Conversely, if L does meet each U_i in finitely many plaques, then L is clearly a finite union of compact sets and hence compact.

Lemma 2.26 *Let R be an open connected saturated subset of M such that each leaf of $\mathcal{F}$ intersecting R is compact with trivial holonomy. Let $a' < a < b < b'$ be real numbers, and $T \colon (a', b') \to M$ a map transversal to $\mathcal{F}$ such that $T((a,b)) \subset R$ and $T(\{a,b\}) \cap R = \emptyset$. Then each leaf $L \subset R$ intersects $T((a,b))$.*

Proof Let W be the union of leaves of $\mathcal{F}$ which intersect $T((a,b))$. We have to show that $W = R$. Since R is connected and W is clearly open and non-empty, it is sufficient to show that W is closed in R.

Let a leaf $L \subset R$ be in the closure of W. By assumption, L is compact with trivial holonomy, so the local Reeb stability theorem implies that the foliation on a small open neighbourhood V of L looks like the product $L \times (-\epsilon, \epsilon)$. Now $T((a,b))$ intersects V, but the end-points $T(a)$ and $T(b)$ are not in V. By transversality it now follows that $T((a,b))$ hits any leaf in V, and in particular L. $\square$

Lemma 2.27 *Let $\gamma \colon \mathbb{R} \to M$ be a curve transverse to $\mathcal{F}$ with the property that $\gamma(\mathbb{R}) \cap L_0 \neq \emptyset$. If L is a leaf of $\mathcal{F}$ with $L \cap \gamma(\mathbb{R}) \neq \emptyset$ then L is diffeomorphic to L_0.*

Proof For any $t \in \mathbb{R}$ denote by L_t the leaf of $\mathcal{F}$ with $\gamma(t) \in L_t$, and put

$$A = \{ t \in \mathbb{R} \mid L_t \text{ is diffeomorphic to } L_0 \} \, .$$

We will show that A is open and closed, hence all of $\mathbb{R}$. If $t \in A$ then note that L_t is compact with trivial holonomy (Lemma 2.25). Thus we can apply the local Reeb stability theorem (Theorem 2.9), so the foliation $\mathcal{F}$ in a neighbourhood of L_t looks like a product $L_t \times (-\epsilon, \epsilon)$. In particular, all the leaves in this neighbourhood are diffeomorphic to L_t. This proves that A is open in $\mathbb{R}$.

Now assume that $[0, s) \subset A$ for some $s > 0$. We will prove that $s \in A$.

(i) First we will show that L_s is compact. Let U_i be a chart from our finite atlas of M. It is enough to show that L_s intersects U_i in a finite number of plaques. Consider the region

$$R = \bigcup_{0 < t < s} L_t \, .$$

If $L_s \subset R$ there is nothing more to prove. Otherwise, each plaque of L_s in U_i lies in the boundary of a different component of $R \cap U_i$ (Figure 2.2). Using the standard transversal section T in the chart U_i, Lemma 2.26 implies that all but possibly two of the components of $R \cap U_i$ intersect any leaf L_t of R. Take any $0 < t < s$. Since L_t is compact, it intersects U_i in finitely many plaques, say m. By the previous observations, L_s can intersect U_i in at most $m + 1$ plaques. Therefore, L_s is compact.

(ii) Observe that since L_t is compact for any $0 \leq t < s$, the leaf L_s must have trivial holonomy 'on the negative side', i.e. on the part $\gamma((s - \epsilon, s])$ of the transversal section $\gamma((s - \epsilon, s + \epsilon))$. Indeed, if the

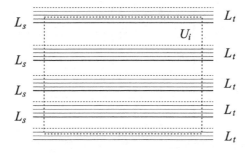

Fig. 2.2. $R \cap U_i$

holonomy of L_s on the negative side were not trivial, then each leaf L_t with $t < s$ close enough to s would hit the chart U_i with $\gamma(s) \in U_i$ in infinitely many plaques; this would contradict the assumption that L_t is compact. Now exactly as for the local Reeb stability theorem, compactness of L_s and finite holonomy on the negative side imply that the foliation $\mathcal{F}$ on $\bigcup_{s-\delta<t\leq s} L_t$ for some small δ looks like the product $L_s \times (-\epsilon, 0]$. In particular, L_s is diffeomorphic to L_t for $s - \delta < t < s$. Thus $s \in A$.

Similarly one proves that $(-s, 0] \subset A$ implies $-s \in A$, and hence we may conclude that A is closed. $\qquad\square$

Lemma 2.28 *There exists a smooth embedding* $\sigma \colon S^1 \to M$ *transverse to the leaves of* $\mathcal{F}$ *which hits each leaf of* $\mathcal{F}$ *exactly once.*

Proof Choose a Riemannian metric on M and let X be a normalized normal field of $\mathcal{F}$ (this exists since $\mathcal{F}$ is transversely orientable). Let γ be an integral curve of X with $\gamma(0) \in L_0$. By Lemma 2.27 all the leaves hit by γ are diffeomorphic to L_0. If γ is periodic with period p, then we take σ' to be $\gamma|_{[0,p]}$ with the natural reparametrization. Otherwise γ is an injective immersion which is not closed. Take a point p in the boundary of $\gamma(\mathbb{R})$, and choose a foliation chart $\varphi \colon U \to \mathbb{R}^{n-1} \times \mathbb{R}$ with $p \in U$. We may also assume that X is tangent to the fibres of $\mathrm{pr}_1 \circ \varphi$. Now γ intersects U in infinitely many segments close to p, and one can easily modify a part of γ inside U to obtain a closed embedded curve σ' transverse to $\mathcal{F}$ (Figure 2.3). Since σ' partially coincides with γ, Lemma 2.27 implies that all the leaves of $\mathcal{F}$ hit by σ' are diffeomorphic to L_0.

Now observe that, by compactness, σ' hits each leaf finitely often. Choose a leaf L which is hit by σ', and let z_0 and z_1 be two consecutive

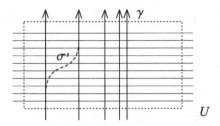

Fig. 2.3. Modification of γ

points on S^1 with $\sigma'(z_0), \sigma'(z_1) \in L$. Choose a simple smooth path β in L from $y = \sigma'(z_1)$ to $x = \sigma'(z_0)$. We can now modify the concatenation of β with the restriction of σ' to the interval between z_0 and z_1 in S^1 to obtain an embedded closed curve σ transversal to $\mathcal{F}$: this modification can be done in a small tubular neighbourhood of $\beta([0,1])$ so that σ hits L exactly once (Figure 2.4). Note that by the same argument as before, all the leaves of $\mathcal{F}$ hit by σ are diffeomorphic to L_0.

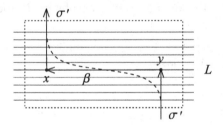

Fig. 2.4. Modification of σ'

For any $z \in S^1$ let L_z be the leaf with $\sigma(z) \in L_z$. The leaf L_z is compact with trivial holonomy, so the foliation of a small saturated open neighbourhood V of L_z looks like the product $L_z \times (-\epsilon, \epsilon)$. The transversality of σ implies that σ intersects V in a finite number of segments, each of them intersecting each leaf in V exactly once. Therefore $\#(L_z \cap \sigma(S^1))$ is a locally constant function of z. Since σ hits L exactly once, we should have $\#(L_z \cap \sigma(S^1)) = 1$ for any $z \in S^1$.

Finally, let $B \subset M$ be the union of leaves of $\mathcal{F}$ which are hit by σ. By transversality of σ it follows that B is open. On the other hand, B is also closed. Indeed, by the local Reeb stability theorem, the union of leaves hit by σ restricted to a small closed interval in S^1 is diffeomorphic to the product of the interval with L_0 and hence compact. By compactness of

S^1 we can thus write B as a finite union of compact sets. This implies that B is compact and hence closed, and connectedness of M gives $B = M$. □

Proof (of Theorem 2.24) Choose $\sigma\colon S^1 \to M$ as in Lemma 2.28. For each $z \in S^1$ denote by L_z the leaf of $\mathcal{F}$ with $\sigma(z) \in L_z$. Define $\phi\colon M \to S^1$ with the condition $x \in L_{\phi(x)}$. □

REMARK. In fact, to prove the theorem it is not necessary to choose σ so that it hits any leaf exactly once. If it hits any leaf exactly p times, $p > 0$, then one can define a p-valued function $\tilde{\phi}$ on M by $\tilde{\phi}(x) = \{z \in S^1 \,|\, x \in L_z\}$. One can then easily see that this p-valued function factors through a p-fold cover of S^1 as a submersion $\phi\colon M \to S^1$ with the required properties. (Recall from complex analysis that a p-valued function on M is a function on a p-fold covering space of M.)

Exercises 2.29 (1) Show that the assumption in the global Reeb stability theorem that the manifold is compact is necessary: construct a foliation of codimension 1 on $\mathbb{R}^3 - \{0\}$ with compact and non-compact leaves and with all the holonomy groups trivial.

(2) Let $\mathcal{F}$ be a foliation of codimension 1 with only compact leaves. Then the holonomy group of any leaf of $\mathcal{F}$ is either trivial or isomorphic to $\mathbb{Z}_2$. If $\mathcal{F}$ is also transversely orientable, then any leaf of $\mathcal{F}$ has trivial holonomy.

(3) Suppose that $\mathcal{F}$ is a transversely orientable codimension 1 foliation of a connected manifold M with only compact leaves. Show that:

   (i) If M is compact, then the leaves of $\mathcal{F}$ are the fibres of a fibre bundle over S^1.

   (ii) If M is non-compact, then the leaves of $\mathcal{F}$ are the fibres of a fibre bundle over $\mathbb{R}$.

In particular, conclude from (ii) that there is no foliation of codimension 1 of the plane $\mathbb{R}^2$ with only compact leaves.

2.6 Thurston's stability theorem

From the proof of the global Reeb stability theorem it is clear that the assumption that the foliation has a compact leaf L_0 with finite fundamental group can be replaced by a weaker one. What we need is a condition that guarantees that any leaf diffeomorphic to L_0 has itself trivial holonomy. This is true, for example, if the fundamental group of L_0 is generated by elements of finite order, by the argument given in the proof of Lemma 2.25. Thurston (1974) showed that in fact it is enough to assume that there are no non-trivial homomorphisms from the fundamental group of the leaf to $\mathbb{R}$, or in other words, that the de Rham cohomology of the leaf is trivial in degree 1. In this section we shall prove his strengthened version of the global stability:

Theorem 2.30 (Global Reeb–Thurston stability) *If $\mathcal{F}$ is a transversely orientable foliation of codimension 1 of a connected compact manifold M which admits a compact leaf L_0 with trivial de Rham cohomology in degree 1, then $\mathcal{F}$ is the foliation given by the fibres of a fibre bundle projection $\phi\colon M \to S^1$. In particular, all the leaves of $\mathcal{F}$ are diffeomorphic to L_0.*

The proof reduces to the arguments already given in the proof of Theorem 2.24, together the following local version of Reeb–Thurston stability.

Theorem 2.31 (Local Reeb–Thurston stability) *Let $\mathcal{F}$ be a foliation of codimension q of a manifold M with a compact leaf L. Then either*

(i) the linear holonomy group of L is non-trivial, or

(ii) the de Rham cohomology of L is non-trivial in degree 1, or

(iii) the holonomy group of L is trivial, and there exist an open saturated neighbourhood V of L in M and a diffeomorphism

$$V \cong L \times \mathbb{R}^q$$

under which the leaves of $\mathcal{F}$ in V correspond to the fibres of the projection $L \times \mathbb{R}^q \to \mathbb{R}^q$.

Before we give the proofs of these two theorems, we need to prove two lemmas related to approximate homomorphisms of groups, and to the topology of the space of germs of maps $(\mathbb{R}^q, 0) \to (\mathbb{R}^q, 0)$, respectively.

Let G be a group, K a subset of G and $\epsilon \geq 0$. A map $c\colon K \to \mathbb{R}$ is a (K, ϵ)-*cocycle* on G if for any $g, h \in K$ with $gh \in K$ we have

$$|c(g) + c(h) - c(gh)| \leq \epsilon .$$

Such a (K, ϵ)-cocycle c is *normalized* on a subset $B \subset K$ if

$$\max_{g \in B} |c(g)| = 1 .$$

Note that $(G, 0)$-cocycles on G are homomorphisms from G to $\mathbb{R}$.

For a subset $B \subset G$ and any $l = 1, 2, \ldots$ we write

$$B^l = \{g_1 g_2 \ldots g_l \mid g_1, g_2, \ldots, g_l \in B\} \,.$$

Furthermore, we write $B^{-1} = \{g^{-1} \mid g \in B\}$. Note that if B is a set of generators of G such that $1 \in B$ and $B^{-1} = B$, then $B = B^1 \subset B^2 \subset \cdots$ and $\bigcup_{l=1}^{\infty} B^l = G$.

Lemma 2.32 *Let G be a group and $B \subset G$ a finite set of generators of G such that $1 \in B$ and $B^{-1} = B$. If for any $\epsilon > 0$ and any $l \geq 1$ there exists a (B^l, ϵ)-cocycle on G which is normalized on B, then there exists a non-trivial homomorphism of groups $G \to \mathbb{R}$.*

Proof For any $l \geq 1$ and $\epsilon > 0$, let $C(l, \epsilon) \subset \mathbb{R}^{B^l}$ be the set of all (B^l, ϵ)-cocycles on G which are normalized on B. This set is closed and bounded in $\mathbb{R}^{B^l}$, hence compact. Furthermore, it is non-empty for $\epsilon > 0$ by hypothesis, and we have $C(l, \epsilon) \subset C(l, \epsilon')$ for any $\epsilon' > \epsilon > 0$. It follows that

$$C(l, 0) = \bigcap_{\epsilon > 0} C(l, \epsilon)$$

is also a non-empty compact subset of $\mathbb{R}^{B^l}$.

Let $r = r_l \colon C(l+1, 0) \to C(l, 0)$ be given by the restriction of $(B^{l+1}, 0)$-cocycles to $B^l \subset B^{l+1}$. Since this map is continuous, it follows that the set $r^{l-1}(C(l, 0))$ is a compact non-empty subset of $C(1, 0)$. We also have $r^l(C(l+1, 0)) \subset r^{l-1}(C(l, 0))$, so

$$C = \bigcap_{l \geq 1} r^{l-1}(C(l, 0))$$

is a non-empty subset of $C(1, 0)$.

Take any $c \in C$. The $(B, 0)$-cocycle $c \in C$ is normalized on B and has an extension to a $(B^l, 0)$-cocyle, which is in fact unique. All these extensions, for $l = 1, 2, \ldots$, now define a homomorphism $G \to \mathbb{R}$, which is non-trivial because c is normalized on B. $\qquad \square$

For any $q \geq 1$, let us denote by $\mathrm{Maps}_0(\mathbb{R}^q)$ the vector space of germs at 0 of smooth maps $f \colon \mathbb{R}^q \to \mathbb{R}^q$ satisfying $f(0) = 0$. The operation of composition in $\mathrm{Maps}_0(\mathbb{R}^q)$ satisfies

$$(v + w) \circ u = v \circ u + w \circ u$$

for any $u, v, w \in \mathrm{Maps}_0(\mathbb{R}^q)$. The differential at 0 gives us a linear map

$$d_0 \colon \mathrm{Maps}_0(\mathbb{R}^q) \longrightarrow \mathrm{Mat}_{q \times q}(\mathbb{R})$$

to the algebra of real $q \times q$ matrices, $d_0(\mathrm{germ}_0 f) = (df)_0$, which satisfies $d_0(u \circ v) = d_0(u)d_0(v)$.

Lemma 2.33 *For any $q \geq 1$ there exists a sequence $(\|\text{-}\|_n)_{n=1}^{\infty}$ of norms on the vector space $\mathrm{Maps}_0(\mathbb{R}^q)$ such that*

$$\limsup_{n \to \infty} \frac{\|u \circ v - u \circ w\|_n}{\|v - w\|_n} \leq \|d_0(u)\|$$

for any $u, v, w \in \mathrm{Maps}_0(\mathbb{R}^q)$ with $v \neq w$.

Proof Let $\mathrm{Maps}(\mathbb{R}^q)$ be the vector space of all smooth maps $f \colon \mathbb{R}^q \to \mathbb{R}^q$ satisfying $f(0) = 0$. There is also a composition operation in $\mathrm{Maps}(\mathbb{R}^q)$, which is preserved by the quotient map

$$\mathrm{germ}_0 \colon \mathrm{Maps}(\mathbb{R}^q) \longrightarrow \mathrm{Maps}_0(\mathbb{R}^q) \,.$$

This linear map is surjective, hence we can choose a linear section α of germ_0, i.e. $\mathrm{germ}_0 \circ \alpha = \mathrm{id}$.

Now for any $n \geq 1$ define a seminorm $\|\text{-}\|_n$ on $\mathrm{Maps}(\mathbb{R}^q)$ by

$$\|f\|_n = \max_{\|x\| \leq \frac{1}{n}} \|f(x)\| \,.$$

By pull-back along the section α, we obtain a norm $\|\text{-}\|_n$ on the vector space $\mathrm{Maps}_0(\mathbb{R}^q)$, i.e.

$$\|u\|_n = \|\alpha(u)\|_n$$

for any $u \in \mathrm{Maps}_0(\mathbb{R}^q)$. We will show that these norms have the property stated in the lemma.

First of all, for any $f, g, h \in \mathrm{Maps}(\mathbb{R}^q)$ and any $x \in \mathbb{R}^q$, the mean value theorem implies that

$$\|f(g(x)) - f(h(x))\| \leq \|g(x) - h(x)\| \max_{y \in S} \|(df)_y\| \,,$$

where S is the segment between $g(x)$ and $h(x)$. Therefore for any $n \geq 1$ we have

$$\|f \circ g - f \circ h\|_n \leq \|g - h\|_n \max_{y \in K} \|(df)_y\| \,,$$

where $K = \{y \in \mathbb{R}^q \,|\, \|y\| \leq \max\{\|g\|_n, \|h\|_n\}\}$. If $\mathrm{germ}_0 g \neq \mathrm{germ}_0 h$ then $\|g - h\|_n > 0$ for any $n \geq 1$, and continuity of the differential df implies

$$\limsup_{n \to \infty} \frac{\|f \circ g - f \circ h\|_n}{\|g - h\|_n} \leq \|d_0(f)\| \,.$$

Now take any $u, v, w \in \mathrm{Maps}_0(\mathbb{R}^q)$ with $v \neq w$. Since we have $\mathrm{germ}_0(\alpha(u \circ z)) = \mathrm{germ}_0(\alpha(u) \circ \alpha(z))$ for $z = v, w$, it follows that

$$
\begin{aligned}
\limsup_{n \to \infty} \frac{\|u \circ v - u \circ w\|_n}{\|v - w\|_n} &= \limsup_{n \to \infty} \frac{\|\alpha(u \circ v) - \alpha(u \circ w)\|_n}{\|\alpha(v) - \alpha(w)\|_n} \\
&= \limsup_{n \to \infty} \frac{\|\alpha(u) \circ \alpha(v) - \alpha(u) \circ \alpha(w)\|_n}{\|\alpha(v) - \alpha(w)\|_n} \\
&\leq \|d_0(\alpha(u))\| \\
&= \|d_0(u)\| .
\end{aligned}
$$

$\square$

Proof (of Theorem 2.31) Note that if $\mathrm{Hol}(L)$ is trivial then the rest of (iii) follows by the local Reeb stability theorem (Theorem 2.9). Assume that the holonomy homomorphism

$$
\mathrm{hol} \colon \pi_1(L) \longrightarrow \mathrm{Diff}_0(\mathbb{R}^q) \subset \mathrm{Maps}_0(\mathbb{R}^q)
$$

is non-trivial, and that its composition with the differential d_0 is trivial. We have to show that this implies $H^1_{\mathrm{dR}}(L, \mathbb{R}) = \mathrm{Hom}(\pi_1(L), \mathbb{R}) \neq 1$.

Since L is compact, the fundamental group of L is finitely generated, so we can choose a finite set of generators $B \subset \pi_1(L)$ with $1 \in B$ and $B^{-1} = B$. By Lemma 2.32 it is sufficient to show that there exists a (B^l, ϵ)-cocycle on $\pi_1(L)$ which is normalized on B, for any $l \geq 1$ and $\epsilon > 0$.

Let $l \geq 1$ and $\epsilon > 0$, and choose $\delta > 0$ to be small enough that $((l-1)\delta + l)\delta \leq \epsilon$. Choose a sequence of norms $(\|\cdot\|_n)_{n=1}^{\infty}$ as in Lemma 2.33. For any $g \in \pi_1(L)$ we have $\|d_0(\mathrm{hol}(g) - \mathrm{id})\| = 0$, and hence for any other $h \in \pi_1(L)$ with $\mathrm{hol}(h) \neq \mathrm{id}$ it follows that

$$
\limsup_{n \to \infty} \frac{\|(\mathrm{hol}(g) - \mathrm{id}) \circ \mathrm{hol}(h) - (\mathrm{hol}(g) - \mathrm{id})\|_n}{\|(\mathrm{hol}(h) - \mathrm{id})\|_n} = 0 .
$$

Since B^l is finite, we can therefore choose $n \geq 1$ so large that

$$
\|(\mathrm{hol}(g) - \mathrm{id}) \circ \mathrm{hol}(h) - (\mathrm{hol}(g) - \mathrm{id})\|_n \leq \|(\mathrm{hol}(h) - \mathrm{id})\|_n \delta \quad (2.1)
$$

for any $g, h \in B^l$.

Now write $M = \max_{g \in B} \|(\mathrm{hol}(g) - \mathrm{id})\|_n$. Then $M > 0$, and we define $\eta \colon B^l \to \mathrm{Maps}_0(\mathbb{R}^q)$ by

$$
\eta(h) = \frac{1}{M}(\mathrm{hol}(h) - \mathrm{id}) .
$$

First we will show that for any $1 \leq k \leq l$ and any $h \in B^k$ we have

$$\|\eta(h)\|_n \leq (k-1)\delta + k \,. \tag{2.2}$$

This is clearly true for $k = 1$, because by definition of η we have $\|\eta(h)\|_n \leq 1$ for any $h \in B$. We shall now proceed by induction on k, so assume that the relation (2.2) holds for some $1 \leq k < l$. Take any $h \in B^{k+1}$, and write $h = h'h''$ for some $h' \in B^k$ and $h'' \in B$. Now we have

$$
\begin{aligned}
\|\eta(h)\|_n &= \frac{1}{M}\|\mathrm{hol}(h) - \mathrm{id}\|_n \\
&= \frac{1}{M}\|\mathrm{hol}(h'h'') - \mathrm{hol}(h'') - \mathrm{hol}(h') + \mathrm{id} \\
&\quad\quad + \mathrm{hol}(h') - \mathrm{id} + \mathrm{hol}(h'') - \mathrm{id}\|_n \\
&\leq \frac{1}{M}\|(\mathrm{hol}(h') - \mathrm{id}) \circ \mathrm{hol}(h'') - (\mathrm{hol}(h') - \mathrm{id})\|_n \\
&\quad\quad + \frac{1}{M}\|\mathrm{hol}(h') - \mathrm{id}\|_n + \frac{1}{M}\|\mathrm{hol}(h'') - \mathrm{id}\|_n \\
&\leq \frac{1}{M}\|\mathrm{hol}(h'') - \mathrm{id}\|_n\delta \\
&\quad\quad + \frac{1}{M}\|\mathrm{hol}(h') - \mathrm{id}\|_n + \frac{1}{M}\|\mathrm{hol}(h'') - \mathrm{id}\|_n \\
&= \|\eta(h'')\|_n\delta + \|\eta(h')\|_n + \|\eta(h'')\|_n \\
&\leq \delta + (k-1)\delta + k + 1 \\
&= k\delta + (k+1) \,.
\end{aligned}
$$

The second inequality here follows from the relation (2.1), while the last is a consequence of the induction hypothesis together with the fact that $\|\eta(h)\|_n \leq 1$ for any $h \in B$.

The relation (2.2) (for $k = l$) now implies that for any $g, h \in B^l$ satisfying $gh \in B^l$ we have

$$
\begin{aligned}
\|\eta(gh) &- \eta(g) - \eta(h)\|_n \\
&= \frac{1}{M}\|\mathrm{hol}(gh) - \mathrm{id} - \mathrm{hol}(g) + \mathrm{id} - \mathrm{hol}(h) + \mathrm{id}\|_n \\
&= \frac{1}{M}\|(\mathrm{hol}(g) - \mathrm{id}) \circ \mathrm{hol}(h) - (\mathrm{hol}(g) - \mathrm{id})\|_n \\
&\leq \frac{1}{M}\|\mathrm{hol}(h) - \mathrm{id}\|_n\delta \\
&= \|\eta(h)\|_n\delta \\
&\leq ((l-1)\delta + l)\delta \leq \epsilon \,.
\end{aligned}
$$

Choose $g \in B$ with $\|\eta(g)\|_n = 1$, and let f be any bounded linear functional on $(\text{Maps}_0(\mathbb{R}^q), \| \cdot \|_n)$ of norm 1 which equals 1 on $\eta(g)$. Then $f \circ \eta$ is a (B^l, ϵ)-cocycle on $\pi_1(L)$ which is normalized on B. $\qquad \square$

Proof (of Theorem 2.30) Let L be a leaf of $\mathcal{F}$ with $H^1_{\text{dR}}(L, \mathbb{R}) = 0$, and consider the sequence of homomorphisms of groups

$$\pi_1(L) \xrightarrow{\text{hol}} \text{Diff}_0(\mathbb{R}) \xrightarrow{d.} \mathbb{R}^+ \xrightarrow{\log} \mathbb{R} \ .$$

Since $H^1_{\text{dR}}(L, \mathbb{R}) = \text{Hom}(\pi_1(L), \mathbb{R}) = 0$, we have $d_0 \circ \text{hol} = 1$, thus the linear holonomy of L is trivial. Theorem 2.31 now implies that the holonomy of L is trivial and that the foliation looks like the product on a neighbourhood of L. With this, we can proceed as in the proof of Theorem 2.24. $\qquad \square$

3

Two classical theorems

In this chapter we present two milestones of the early theory of foliations, namely the theorems of Haefliger and of Novikov. Both theorems concern foliations of codimension 1 on three-dimensional manifolds.

Haefliger's theorem dates from the late 1950s, and concerns the problem of constructing codimension 1 foliations on 3-manifolds. One version asserts that if a compact three-dimensional manifold carries an analytic foliation of codimension 1, then this manifold must have infinite fundamental group. Thus, such a foliation cannot exist on the 3-sphere, for example. We will present a detailed proof, which is close to Haefliger's original argument, and which involves various notions of independent interest. The first of these is that of a Morse function into a manifold carrying a codimension 1 foliation, which we discuss in Subsection 3.1.2, after having reviewed the classical theory of Morse functions into the line. The other is that of foliations with isolated singularities on a two-dimensional disk, to be discussed in Subsection 3.1.3. These singular foliations arise by pull-back along a Morse function from the disk into a given 3-manifold equipped with a codimension 1 foliation.

Novikov's theorem dates from the 1960s, and concerns the existence of compact leaves. Explicitly, it states that any (smooth) transversely orientable codimension 1 foliation of a compact manifold with finite fundamental group must have a compact leaf. Moreover, a closer analysis reveals that this compact leaf must be a torus, and that inside this torus, the given foliation looks exactly like the Reeb foliation discussed in Example 1.1 (5). We will also present a detailed proof of this result of Novikov's. As the reader will see, the proof is in part based on some of the techniques developed by Haefliger for the proof of his theorem. An additional important notion involved in the proof of Novikov's theorem is that of a 'vanishing cycle'. Such a cycle is a non-contractible loop inside a leaf, which becomes contractible as soon as you slide it into nearby leaves; in other words, it represents an element of the fundamental group of a leaf, which vanishes in the fundamental groups of nearby leaves. Such vanishing cycles will be discussed in Subsection 3.2.1. They play an important role in many parts of foliation theory.

3.1 Haefliger's theorem

The aim of this section is to prove the following theorem of Haefliger.

Theorem 3.1 (Haefliger) *There are no analytic codimension 1 foliations on S^3.*

For a foliation of codimension 1, the holonomy of a loop is (represented by) a germ of a diffeomorphism $g\colon (\mathbb{R}, 0) \to (\mathbb{R}, 0)$. Call such a germ *one-sided* if g restricts to the identity germ either on $((-\infty, 0], 0)$ or on $([0, \infty), 0)$. We say that the foliation is without (non-trivial) one-sided holonomy if any one-sided holonomy germ of the foliation is trivial. Clearly, any analytic foliation of codimension 1 is without one-sided holonomy. The proof of the theorem above will in fact show the following stronger statement.

Theorem 3.2 *Let $(M, \mathcal{F})$ be a codimension 1 foliated manifold without one-sided holonomy. Then every loop in M transverse to $\mathcal{F}$ represents an element of $\pi_1(M)$ of infinite order.*

Recall from Section 2.5 that on a compact manifold M there are always plenty of such transverse loops. The proof there used that $(M, \mathcal{F})$ is transversely orientable. For general $(M, \mathcal{F})$, construct first the *transverse orientation cover* $t\colon \tilde{M} \to M$ on which $\mathcal{F}$ pulls back to a transversely orientable foliation $t^*(\mathcal{F})$. Then any transverse loop $\tilde{\alpha}$ in $(\tilde{M}, t^*(\mathcal{F}))$ projects to the transverse loop $t \circ \tilde{\alpha}$ in $(M, \mathcal{F})$.

In particular, it follows that a compact manifold with finite fundamental group cannot have an analytic codimension 1 foliation.

One possible proof of Haefliger's theorem 3.2 proceeds as follows. Suppose to the contrary that α is a transverse loop in $(M, \mathcal{F})$ with finite order in $\pi_1(M)$, say k. By replacing α by α^k, we may assume that $[\alpha] = 1$ in $\pi_1(M)$. Thus there exists an extension of α to a map $H\colon D \to M$, where D is the unit disk in $\mathbb{R}^2$. Now $\mathcal{F}$ pulls back to a 'foliation with singularities' $H^*(\mathcal{F})$ of D. Note that since $H|_{S^1} = \alpha$, this 'foliation' is transverse to the boundary $S^1 = \partial D$ and all the singularities are in the interior of D. Then we use Morse functions to deform H a little, so that these singularities have a 'normal form'. From this, we will conclude that there exists a loop $\beta\colon S^1 \to D$ whose composition with H represents a loop with non-trivial one-sided holonomy in $(M, \mathcal{F})$. In the following subsections we will work out the details of this plan.

3.1.1 Review of Morse functions

We will first recall some basic facts from Morse theory. For proofs, see Milnor (1963), Golubitsky–Guillemin (1973) and Guillemin–Pollack (1974). Consider a smooth function $f\colon \mathbb{R}^n \to \mathbb{R}$. Its derivative at $p \in \mathbb{R}^n$ is a linear map $(df)_p\colon \mathbb{R}^n \to \mathbb{R}$. A *critical point* or *singularity* of f is a point $p \in \mathbb{R}^n$ with $(df)_p = 0$, or equivalently $\frac{\partial f}{\partial x_i}(p) = 0$ for $i = 1, \ldots, n$. The value of f at a critical point is called a *critical value* of f.

At a critical point p, the second partial derivatives form a symmetric $n \times n$ matrix

$$H(f)_p = \left(\frac{\partial^2 f}{\partial x_i \partial x_j}(p) \right)_{i,j} ,$$

which is called the Hessian matrix of f at p. One says that p is a non-degenerate singularity of f if $\det H(f)_p \neq 0$. This is equivalent to the associated Hessian quadratic form on $\mathbb{R}^n$

$$\bar{H}(f)_p(u) = \langle H(f)_p u, u \rangle$$

being non-degenerate. Observe that non-degenerate singularities are automatically isolated.

For a function $f\colon M \to \mathbb{R}$ on a manifold M of dimension n, one can use the local coordinates to give the same definition of singularity and non-degenerate singularity (called also *Morse* singularity). The definition is independent of the coordinates (exercise). A *Morse function* on M is a smooth function $f\colon M \to \mathbb{R}$ all of whose singularities are non-degenerate.

Lemma 3.3 (Morse lemma) *Let $f\colon M \to \mathbb{R}$ be a smooth function on a manifold M of dimension n. For any non-degenerate singularity p of f there are local coordinates $(x_1, \ldots, x_n)\colon U \to \mathbb{R}^n$, defined on a neighbourhood U of p and with $x_i(p) = 0$, on which f has the form*

$$f = f(p) - x_1^2 - \cdots - x_i^2 + x_{i+1}^2 + \cdots + x_n^2 .$$

REMARK. Note that $H(f)_p = \mathrm{diag}(-2, \ldots, -2, 2, \ldots, 2)$ in these coordinates, where -2 appears on the diagonal i times. The number i, called the (Morse) *index* of f at p and denoted by $\mathrm{Index}(f,p)$, is independent of the choice of coordinates.

For a smooth function $f\colon M \to \mathbb{R}$, a metric on M enables one to write the derivative $(df)_p\colon T_p(M) \to \mathbb{R}$ as the dual of a tangent vector $\mathrm{grad}(f)_p \in T_p(M)$, i.e.

$$(df)_p(v) = \langle v, \mathrm{grad}(f)_p \rangle\,.$$

This defines a vector field $\mathrm{grad}(f)$ on M.

Exercise 3.4 Prove that the Morse index of $f\colon M \to \mathbb{R}$ and the index of the vector field $\mathrm{grad}(f)$ at a non-degenerate singularity $p \in M$ of f are related by

$$(-1)^{\mathrm{Index}(f,p)} = \mathrm{Index}_p(\mathrm{grad}(f))\,.$$

Recall now that the vector space $C^\infty(M,\mathbb{R})$ of smooth real-valued functions on a manifold M is equipped with the C^∞-topology: it is the topology of uniform convergence on compact sets of functions and all their higher derivatives. The subbasic neighbourhoods of $f \in C^\infty(M,\mathbb{R})$ are

$$B_{\epsilon,K,\alpha}(f) = \{ g \in C^\infty(M,\mathbb{R}) \mid \|D^\beta(f-g)|_K\| < \epsilon \text{ for all } \beta \leq \alpha \}\,,$$

where $\epsilon > 0$, K is a compact subset contained in some chart of M, $\alpha = (\alpha_1, \ldots, \alpha_n)$ is a multi-index, and $\beta \leq \alpha$ means that β is a multi-index with $\beta_i \leq \alpha_i$ for all i. Finally, D^β is the partial derivative $(\frac{\partial}{\partial x_1})^{\beta_1} \cdots (\frac{\partial}{\partial x_n})^{\beta_n}$, in the local coordinates, and $\|\text{-}\|$ denotes the maximum norm.

Note that there is a similar topology on the space $C^\infty(M,N)$ of smooth functions between smooth manifolds M and N, defined e.g. by an embedding of N into some $\mathbb{R}^k$ and defining $C^\infty(M,N)$ as a subspace of $C^\infty(M,\mathbb{R})^k$.

Exercise 3.5 Think a bit about this topology. Show e.g. that $C^\infty(M,\mathbb{R})$ is a topological algebra, i.e. addition and multiplication are continuous; show that for a smooth function $f\colon \mathbb{R} \times M \to \mathbb{R}$, the 'transposed' map $F\colon \mathbb{R} \to C^\infty(M,\mathbb{R})$ is continuous; also show that the composition $C^\infty(N,\mathbb{R}) \times C^\infty(M,N) \to C^\infty(M,\mathbb{R})$ is continuous.

We will use the following application of Sard's theorem; for a proof, see e.g. Guillemin–Pollack (1974).

Theorem 3.6 *The Morse functions on a manifold M form a dense open subset of $C^\infty(M, \mathbb{R})$.*

This result has the following easy variations.

(a) The set of Morse functions with distinct critical values is dense and open in $C^\infty(M, \mathbb{R})$.

(b) Suppose that $U \subset \bar{U} \subset V \subset M$ and $f \in C^\infty(M, \mathbb{R})$. Then any neighbourhood of f contains $g \in C^\infty(M, \mathbb{R})$ with $g|_U$ Morse and $g|_{M-V} = f|_{M-V}$.

To see that Theorem 3.6 implies (b), fix a smooth function $\phi \colon M \to \mathbb{R}$ with $\phi|_U = 1$ and $\phi|_{M-V} = 0$. For any $h \in C^\infty(M, \mathbb{R})$ define

$$G(h) = \phi h + (1 - \phi)f .$$

Note that $G(h)|_U = h|_U$, $G(h)|_{M-V} = f|_{M-V}$ and $G(f) = f$. By continuity, $G(h)$ is arbitrarily close to f if h is sufficiently close to f. So if we use Theorem 3.6 to choose h Morse and close to f, we find $g = G(h)$ as required. We leave the proof of (a) as an exercise.

3.1.2 Morse functions into codimension 1 foliations

Let $(N, \mathcal{F})$ be a codimension 1 foliated manifold, and M any manifold. A *Morse function* from M to $(N, \mathcal{F})$ is a smooth function $f \colon M \to N$ for which there exists a Haefliger cocycle (W_i, s_i, γ_{ij}) on N defining $\mathcal{F}$ so that

$$s_i \circ f|_{f^{-1}(W_i)} \colon f^{-1}(W_i) \longrightarrow \mathbb{R}$$

is Morse, for any i. A point $p \in M$ is a *singularity* of such a Morse function f if it is a singularity of $s_i \circ f|_{f^{-1}(W_i)}$, $f(p) \in W_i$.

Proposition 3.7 *For a compact manifold M, the Morse functions from M to $(N, \mathcal{F})$ form a dense open subset of $C^\infty(M, N)$.*

Proof Let $f \in C^\infty(M, N)$ and let (W_i, s_i, γ_{ij}) be a Haefliger cocycle on N representing $\mathcal{F}$; we assume also that each W_i is the domain of a surjective foliation chart φ_i with $s_i = \mathrm{pr}_2 \circ \varphi_i$, and put $t_i = \mathrm{pr}_1 \circ \varphi_i$. Choose a finite cover of M by charts $(V_k)_{k=1}^m$ such that $f(V_k) \subset W_{i_k}$. Choose a refinement (U_k) with $\bar{U}_k \subset V_k$. We will gradually modify f by considering one U_k at a time.

For U_1, choose $h \colon f^{-1}(W_{i_1}) \to \mathbb{R}$ close to $s_{i_1} \circ f$ with $h|_{U_1}$ Morse, and such that $h|_{f^{-1}(W_{i_1})-V_1} = s_{i_1} \circ f|_{f^{-1}(W_{i_1})-V_1}$. Then let $g_1: M \to N$ be given by

$$g_1(x) = \begin{cases} f(x), & x \notin V_1, \\ \varphi_{i_1}^{-1}(t_{i_1}(f(x)), h(x)), & x \in V_1. \end{cases}$$

For U_2, find a similar g_2 close to g_1 so that $g_2 = g_1$ outside V_2, $s_{i_2} \circ g_2$ is Morse on U_2, and g_2 is so close to g_1 that $s_{i_1} \circ g_2$ is still Morse on U_1. Continue in this way, and finish after m steps. □

There are variations of this proposition, exactly like the variations (a) and (b) of Theorem 3.6 in Subsection 3.1.1. In particular:

(a) The set of Morse functions from M to $(N, \mathcal{F})$ whose critical values lie on distinct leaves is dense in $C^\infty(M, N)$.

(Exercise: Formulate and prove the analogue of (b).)

Assume now that $\mathcal{F}$ is a transversely oriented foliation on N and that $f: M \to (N, \mathcal{F})$ is a Morse function. Let $p \in M$ be a singularity of f. Then we can associate to p the *Morse index* Index(f, p) of f at p as follows. Choose any $s_i: W_i \to \mathbb{R}$ with $f(p) \in W_i$ (from a Haefliger cocycle defining $\mathcal{F}$) compatible with the transverse orientation of $\mathcal{F}$. Then define

$$\text{Index}(f, p) = \text{Index}(s_i \circ f|_{f^{-1}(W_i)}, p).$$

Note that this definition is independent of the choice of s_i.

Proposition 3.8 (Poincaré–Hopf) *Let M be a compact manifold, $(N, \mathcal{F})$ a transversely oriented foliated manifold of codimension 1 and $f: M \to (N, \mathcal{F})$ a Morse function. Then*

$$\sum_p (-1)^{\text{Index}(f,p)} = \chi(M),$$

where the sum is over all the singularities p of f.

Proof Choose a finite number of submersions $s_i: W_i \to \mathbb{R}$, $i = 1, 2, \ldots, k$, from a Haefliger cocycle on N representing $\mathcal{F}$ compatible with the transverse orientation, such that $(V_i = f^{-1}(W_i))_{i=1}^k$ covers M. We can assume that each V_i contains at most one singularity of f. On each V_i we have the vector field $\text{grad}(s_i \circ f)$, computed with respect to any fixed Riemannian metric on M. By using a partition of unity, we can glue these vector fields together to obtain a global vector field X. We can make sure that around any singularity $p \in V_i$, the field X coincides with $\operatorname{grad}(s_i \circ f)$. Now the lemma follows from Exercise 3.4 and the classical Poincaré–Hopf theorem. □

REMARK. The proposition holds true also if M is a manifold with boundary, provided that the 'gradient' of the Morse function f, i.e. the field X from the proof above, can be chosen so that it points outwards at every boundary point of M.

3.1.3 Proof of Haefliger's theorem

We will now prove Theorem 3.2, and proceed as in the outline given at the beginning of Section 3.1. Thus $\mathcal{F}$ is a codimension 1 foliation on a manifold M and $\alpha \colon S^1 \to M$ a closed transversal curve. We will show that the assumption that $[\alpha] = 1$ in $\pi_1(M)$ leads to the existence of a loop in a leaf of $\mathcal{F}$ with non-trivial one-sided holonomy.

Since α is homotopic to zero, it can be extended to a map on the disk

$$H \colon D \longrightarrow M .$$

We may assume that H is smooth (see Bott–Tu (1982)). Notice that since α is transverse to $\mathcal{F}$, H is transverse to $\mathcal{F}$ near the boundary of D. By Proposition 3.7 and its two variants we can thus modify H slightly, to obtain a map

$$K \colon D \longrightarrow M$$

which agrees with H near the boundary, is Morse as a map into $(M, \mathcal{F})$ and has critical values on distinct leaves.

Let $p_1, \ldots, p_k$ be the critical points of K. Outside $p_1, \ldots, p_k$, the map K is transverse to the leaves of $\mathcal{F}$, so $\mathcal{F}$ pulls back to a foliation $\mathcal{F}' = K^*(\mathcal{F})$ of $D - \{p_1, \ldots, p_k\}$, with leaves transverse to the boundary. On D itself, we can interpret $\mathcal{F}'$ as a foliation with singularities. If (W_i, s_i, γ_{ij}) is a Haefliger cocycle representing $\mathcal{F}$, then the connected components of the fibres of $s_i \circ K|_{K^{-1}(W_i)}$ locally define $\mathcal{F}'$.

Now by the Morse lemma, these singularities p_m are of three possible forms, according to whether their index is 0, 1 or 2. For index 0 or 2, $s_i \circ K$ has a local minimum or maximum in p_m (where i is such that $p_m \in K^{-1}(W_i)$), and the foliation $\mathcal{F}'$ near p_m looks like a family of concentric circles (a centre, or an elliptic singularity at p_m), i.e. like the level-sets of the function $x^2 + y^2$ on $\mathbb{R}^2$. For index 1, the point p_m is a saddle point of $s_i \circ K$ (a saddle singularity at p_m), and the foliation near p_m looks like the level-sets of the function $x^2 - y^2$ on $\mathbb{R}^2$ (Figure 3.1).

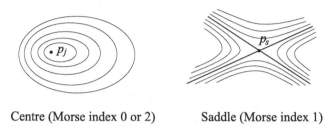

Centre (Morse index 0 or 2)   Saddle (Morse index 1)

Fig. 3.1. Centre and saddle

Lemma 3.9 *The number of elliptic singularities is* 1 *bigger than the number of saddle singularities.*

Proof If $\mathcal{F}$ is transversely orientable, Proposition 3.8 yields

$$1 = \chi(D) = \#\text{centres} - \#\text{saddles} .$$

Otherwise, we replace $(M, \mathcal{F})$ with the double transverse orientation cover $(\tilde{M}, \tilde{\mathcal{F}})$ which is transversely orientable. Since D is simply connected, the map K lifts to a Morse map $\tilde{K} \colon D \to (\tilde{M}, \tilde{\mathcal{F}})$, and we apply Proposition 3.8 to $\tilde{K}$. Finally, note that the centres, respectively saddles, of K are exactly the centres, respectively saddles, of $\tilde{K}$. $\square$

Now consider a centre p_m. The union of the family of all concentric circles around p_m has a boundary curve Γ. This curve Γ is disjoint from ∂D, since $\mathcal{F}'$ is transverse to ∂D. (Indeed, if Γ hits the boundary, there is the 'first' concentric circle which 'touches' the boundary, contradicting transversality.) This curve Γ bounds a disk, and is again a simple closed curve, and a leaf of $\mathcal{F}'$. If Γ does not have a singularity, then Γ is the 'last' concentric circle around p_m, and the leaves of $\mathcal{F}'$ outside Γ must spiral around Γ (Figure 3.2). Thus, Γ evidently has trivial holonomy on the inside and non-trivial holonomy on the outside. The same is true for $K(\Gamma)$.

This proves the theorem in case there exists a centre p_m with a 'boundary concentric circle' Γ without singularities.

If, on the other hand, the boundary curve Γ has a singularity (say p_s), it can have only one (since the singularities lie on distinct leaves) and this one must be a saddle (Figure 3.3). Suppose then that this is true

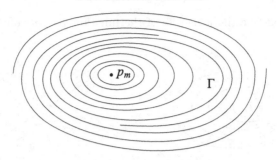

Fig. 3.2. $\mathcal{F}'$ spiraling around Γ

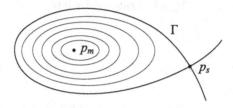

Fig. 3.3. A saddle on Γ

for each centre. By Lemma 3.9, we find that there will be two centres, say p_1 and p_2 with boundary curves Γ_1 and Γ_2, which share the same saddle. There are two typical situations which can occur (Figure 3.4):

Case 1. $\Gamma_1 \cap \Gamma_2 = \{p_s\}$ (then put $\Gamma = \Gamma_1 \cup \Gamma_2$).

Case 2. $\Gamma_1 \subset \Gamma_2$ (then let $\Gamma = (\Gamma_2 - \Gamma_1) \cup \{p_s\}$).

We will prove the theorem in each of these two cases, and then argue that these are the only possibilities.

Case 1. Clearly Γ_1 and Γ_2 have trivial holonomy on the inside. Since M has no non-trivial one-sided holonomy, Γ_1 and Γ_2 have trivial holonomy when mapped to M. Then the composite curve Γ has trivial holonomy in M as well. But this means that for the singular foliation $\mathcal{F}'$ on D, there are concentric circles outside Γ. Then we could replace the inside of Γ by concentric circles and finish the proof by induction on the number of singularities (Figure 3.5).

Case 2. Completely analogous: Γ_1 and Γ_2 have trivial one-sided holonomy, hence so does Γ. Thus Γ has trivial holonomy, and there are circles again on the outside of Γ_2.

Finally, it remains to be shown that these are the only two cases. This can be seen by study of the leaves near the saddle, where the two

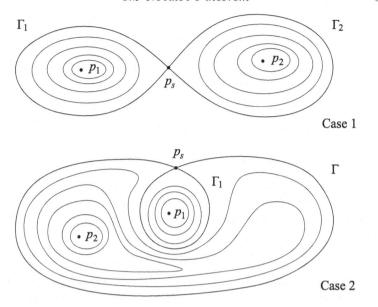

Fig. 3.4. The two typical situations

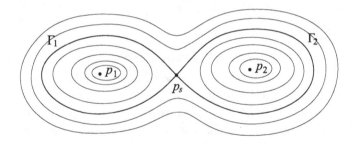

Fig. 3.5. Concentric circles around Γ

families of concentric circles meet at the saddle in only two possible different ways (Figure 3.6).

3.2 Novikov's theorem

In this section M is a compact manifold equipped with a transversely oriented foliation $\mathcal{F}$ of codimension 1. Although some of the tools being developed work in arbitrary dimensions, the main results concern the case where the dimension of M is 3.

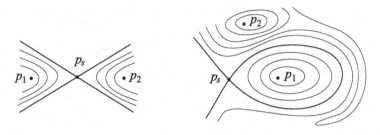

Fig. 3.6. The leaves near the saddle

Theorem 3.10 (Novikov) *Let $\mathcal{F}$ be a codimension 1 transversely oriented foliation of a compact manifold M of dimension 3 with finite fundamental group. Then $\mathcal{F}$ has a compact leaf.*

In addition, it will be shown that if M is also orientable then the compact leaf constructed for Theorem 3.10 is a torus, inside of which the foliation is the Reeb foliation. One says that $(M, \mathcal{F})$ has a *Reeb component*.

The main tool is the method of 'vanishing cycles', a refinement of the method used to prove Haefliger's theorem.

3.2.1 Vanishing cycles

Recall that M is assumed to be a compact manifold and $\mathcal{F}$ a transversely oriented foliation of M of codimension 1.

Let L_0 be a leaf of $(M, \mathcal{F})$. A closed curve $\alpha_0 \colon [0,1] \to L_0$ is said to be a *vanishing cycle* if α_0 is not contractible in L_0, and, for some $\epsilon > 0$, α_0 can be extended to a smooth family

$$\alpha \colon [0,1] \times [0, \epsilon) \longrightarrow M$$

of closed curves $\alpha_t = \alpha(\,\text{-}\,, t)$ such that for each $0 < t < \epsilon$ the curve α_t is a closed curve in a leaf L_t of $\mathcal{F}$ and is contractible in L_t, and such that for each fixed $s \in [0,1]$ the segment $\alpha(s, \,\text{-}\,) \colon [0, \epsilon) \to M$ is transverse to $\mathcal{F}$. One says that α_0 is a *positive (negative)* vanishing cycle if these segments $\alpha(s, \,\text{-}\,)$ are positive (negative) with respect to the transverse orientation of $\mathcal{F}$. Observe that such an α_0 has trivial positive (negative) holonomy.

Example 3.11 Consider the Reeb foliation on the solid torus, and let α_0 represent the standard generator of the boundary torus with trivial holonomy. Then α_0 represents a trivial element of the fundamental group of the solid torus, and a suitable beginning of a contraction of α_0 defines a vanishing cycle (Figure 3.7).

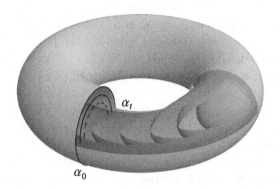

Fig. 3.7. A vanishing cycle of the Reeb foliation

Remarks 3.12 (1) Let $\alpha_0 \colon [0,1] \to L_0$ be a positive vanishing cycle, and put $\alpha_0(0) = \alpha_0(1) = x_0$. Write α_t, $0 \le t < \epsilon$, for its extension as above. Now let $K \subset L_0$ be a compact set containing the image of α_0 (e.g. $K = \alpha_0([0,1])$). Choose a smooth function

$$T \colon K \times (-\delta, \delta) \longrightarrow M$$

such that $T(x, 0) = x$ and $T_x = T(x, (-\delta, \delta))$ is a transversal section of $(M, \mathcal{F})$. We can view this as the choice of a transversal section T_x at x for each $x \in K$, smoothly varying in x. This can be done, for example, by taking the flow of a normal vector field locally on the compact K. We can take $\delta > 0$ small enough so that T is injective. As in the proof of the local Reeb stability in Section 2.3, for $\epsilon > 0$ sufficiently small the holonomy transformations along initial segments of α_0 define a map

$$B \colon [0,1] \times [0, \epsilon) \longrightarrow M$$

by

$$B(s, t) = \mathrm{hol}(\alpha_0|_{[0,s]})(\tau_{x_0}(t)) \,,$$

where $\tau_{x_0} \colon [0, \epsilon) \to T_{x_0}$ is a fixed positive parametrization of a part of the transversal section T_{x_0}, with $\tau_{x_0}(0) = x_0$. Thus $B(s, 0) = \alpha_0(s)$. Since α_0 has trivial positive holonomy, for $\epsilon > 0$ sufficiently small it will be the case that each $B(\,\text{-}\,, t)$ is a closed curve in its leaf.

Suppose that we choose T_{x_0} to be in the image of α, with the evident parametrization $\tau_{x_0}(t) = \alpha_t(0)$. Then $B(\,\text{-}\,,t)$ and α_t lie in the same leaf L_t. For t sufficiently small, $B(\,\text{-}\,,t)$ and α_t are homotopic in L_t, by Exercise 2.3 (3).

This shows that the family of curves $B_t = B(\,\text{-}\,,t)$, $0 \le t < \epsilon$, is also a 'witness' of the fact that α_0 is a vanishing cycle. Notice that this extension (B_t) of the vanishing cycle has the property that whenever transversal sections $B(s, [0, \epsilon))$ and $B(s', [0, \epsilon))$ intersect each other, $\alpha_0(s) = \alpha_0(s')$. We will use this property in the proof of Theorem 3.19 below.

In other words, being a vanishing cycle is a property of α_0 (rather than an extra structure), and a 'witness' can always be found by holonomy: it only depends on the choice of transversal sections.

(2) The property of being a positive (or negative) vanishing cycle only depends on the (free) homotopy class of α_0 (in L_0). This is an easy consequence of the remark (1). Explicitly, suppose $H \colon [0, 1] \times [0, 1] \to L_0$ is a homotopy between loops α_0 and β_0 in L_0. Thus $H(u, \,\text{-}\,)$ is a closed curve in L_0, and $H(0, \,\text{-}\,) = \alpha_0$ while $H(1, \,\text{-}\,) = \beta_0$. Choose a transversal section T_x at x for each x in the image K of H, smoothly varying in x as in the remark (1). We can then lift H by holonomy to a family

$$B \colon [0, 1] \times [0, 1] \times [0, \epsilon) \longrightarrow M$$

by

$$B(u, s, t) = \text{hol}(H(u, \,\text{-}\,)|_{[0,s]})(\tau_{H(u,0)}(t)) \, .$$

Thus $B(u, \,\text{-}\,, \,\text{-}\,)$ is a family of curves $(B_{u,t} = B(u, \,\text{-}\,, t))_{0 \le t < \epsilon}$ extending the curve $B_{u,0} = H(u, \,\text{-}\,)$, for any u. For a fixed t, $B_{u,t}$ defines a homotopy from $B_{0,t}$ to $B_{1,t}$ inside a single leaf if we choose the parametrization $\tau_{H(u,0)}(t)$ suitably, e.g. $\tau_{H(u,0)}(t) = \text{hol}(H(\,\text{-}\,, 0)|_{[0,u]})(\tau_{x_0}(t))$. This shows that $B_{0,0} = \alpha_0$ is a vanishing cycle if and only if $B_{1,0} = \beta_0$ is.

We now prove the existence of vanishing cycles, by exactly the same methods as used in Section 3.1.

Theorem 3.13 *If $\pi_1(M)$ is finite, there exists a vanishing cycle.*

Proof As in Section 3.1, there exists a closed curve α transversal to $\mathcal{F}$. We may assume that α represents the unit element of $\pi_1(M)$, and we can extend α to a Morse map $K \colon D \to M$, defining a foliation with (Morse) singularities of D. Recall that the singularities are centres or saddles, and that there are (one) more centres than saddles.

Consider a centre p. The concentric circles around p define closed curves in the leaves of $\mathcal{F}$, and close to p these curves are contractible in their leaves (they map into a chart around $K(p)$). The family of all these 'contractible' curves defines an open disk around p. Suppose there is a first concentric circle which is not contractible in its leaf. Then this is evidently a vanishing cycle.

If no such exists before we reach the boundary of all the contractible circles around p, consider their boundary Γ. As in Section 3.1, if Γ has no singularities, it is a concentric circle with trivial holonomy on one side and non-trivial holonomy on the other. Thus Γ defines a vanishing cycle.

Suppose we have not found a vanishing cycle in this way. Then each centre p_i is surrounded by 'contractible' concentric circles, and the boundary curve Γ_i contains a saddle singularity. Now at least one saddle p_s must be associated to two centres p_1 and p_2 in this way, so we have one of the following two cases (Subsection 3.1.3, Figure 3.4), as in the proof of Haefliger's theorem (Subsection 3.1.3).

Case 1. $\Gamma_1 \cap \Gamma_2 = \{p_s\}$.

Case 2. $\Gamma_1 \subset \Gamma_2$.

In Case 1, if both Γ_1 and Γ_2 are contractible when mapped to their common leaf in M, they have trivial holonomy, hence so has their composite $\Gamma = \Gamma_1 \cup \Gamma_2$. Thus Γ has 'concentric circles' on its outside, and these are contractible in their leaves when close to Γ. Now proceed as before, to find the first non-contractible closed curve outside Γ, etc. For Case 2, the argument is similar. $\qquad\qquad\qquad\qquad\qquad\qquad\quad\square$

Recall that the global Reeb stability theorem (Section 2.5) gives foliations defined by proper submersions, and these cannot have vanishing cycles. Thus, the preceding theorem contradicts the conclusion of the global Reeb stability theorem. Hence:

Corollary 3.14 *If $\pi_1(M)$ is finite, then any compact leaf of $\mathcal{F}$ must have infinite fundamental group.*

If the dimension of M is 3, it follows that the compact leaves are surfaces which have the plane $\mathbb{R}^2$ as their universal cover.

Consider a vanishing cycle α_0. Of course, α_0 can have self-intersections, but by perturbing α_0 slightly inside the leaf we can assume that these are just double points, and that there are finitely many such. Let α_t, $0 \leq t < \epsilon$, be an extension which witnesses that α_0 is a (say, positive) vanishing cycle, defined by holonomy as in the remark above. Then any self-intersection of α_t maps down (along the transversal section) to a self-intersection of α_0.

The vanishing cycle α_0 is called *simple* if, for ϵ small enough, it holds for all $0 < t < \epsilon$ that α_t lifts to a simple closed curve $\tilde{\alpha}_t$ in the universal cover $\tilde{L}_t$ of L_t.

This definition does not depend on the choice of the extension (α_t). To understand this definition, suppose that α_0 has a self-intersection so that it can be written as the composition of two closed curves β_0 and γ_0. It could be that one of β_0, γ_0 is also a vanishing cycle (in fact, one of them is if and only if the other is: look at the one-sided holonomy of α_0). This means in particular that β_0 and γ_0 have trivial one-sided holonomy, so that α_t decomposes as $\beta_t \cup \gamma_t$ for all small t, and that β_t and γ_t are both contractible in their leaves (so that $\tilde{\alpha}_t$ still has the 'same' intersection).

Suppose, on the other hand, that β_0, γ_0 are not vanishing cycles. Then, if the self-intersection 'persists' for $t > 0$, so that we can write $\alpha_t = \beta_t \cup \gamma_t$, then for any $t > 0$ there exists t' with $0 < t' < t$ so that $\beta_{t'}$, $\gamma_{t'}$ are not contractible in their leaf. Now α_0 is simple if, in this situation, there exists $t > 0$ so small that for any t' with $0 < t' < t$ the curves $\beta_{t'}$ and $\gamma_{t'}$ are not contractible in their leaf.

Thus, to say that α_0 is simple is a bit sharper than saying that α_0 cannot be written as the composition of two vanishing cycles.

Proposition 3.15 *Suppose α_0 is a positive vanishing cycle, with an extension (α_t) as above. Then there exist arbitrarily small $t > 0$ such that L_t contains a simple vanishing cycle.*

Proof We will prove this by induction on the number of self-intersections of α_0. Assume that α_0 is not simple. Then α_0 has a self-intersection, so write $\alpha_0 = \beta_0 \cup \gamma_0$. By the assumption that α_0 is not simple it follows that there are arbitrarily small t' so that $\alpha_{t'} = \beta_{t'} \cup \gamma_{t'}$ and $\beta_{t'}$ and $\gamma_{t'}$ are contractible in their leaf $L_{t'}$. Thus, for the open set

$$C = \{t' \in (0, \epsilon) \mid \alpha_{t'} = \beta_{t'} \cup \gamma_{t'}, \ \beta_{t'} \text{ and } \gamma_{t'} \text{ contractible in } L_{t'}\}$$

we have $0 \in \bar{C}$. If either β_0 or γ_0 is a vanishing cycle, we are done by induction. If not, it follows that $0 \in \overline{((0, \epsilon) - C)}$. Then take $\delta \notin C$ so that $(\delta, \delta + \epsilon) \subset C$ for some $\epsilon > 0$. Now β_δ is a vanishing cycle with fewer self-intersections. $\square$

3.2.2 Existence of a compact leaf

Recall that M is a compact manifold and $\mathcal{F}$ a transversely oriented foliation of M of codimension 1. From now on we shall in addition assume that M is of dimension 3. Moreover, we shall assume, without loss of generality for the proof of Theorem 3.10, that M is connected and orientable. The existence of a compact leaf will now be proved by contradiction, using the following lemma.

Lemma 3.16 *Let L be a non-compact leaf of $\mathcal{F}$, and let $p \in L$. Then there exists a transversal closed curve α passing through p.*

Proof We first show that there exists a transversal closed curve which meets L, and then modify it to pass through p.

Since L is not compact, one can choose a surjective foliation chart $\varphi \colon U \to \mathbb{R}^3$ such that L intersects U in infinitely many plaques. Let σ be a transversal segment from x to y in U connecting two of these plaques, say in the positive direction. Now choose a path β in L from y to x, and modify $\beta \cup \sigma$ slightly inside a tubular neighbourhood around the image of β so that it becomes transversal to $\mathcal{F}$ (Figure 3.8). This gives a transversal closed curve α which intersects L, say in a point q.

Finally, choose a path τ in L from p to q, and modify $\alpha \cup \tau \cup \tau^{-1}$ slightly inside a tubular neighbourhood around the image of τ so that it is transversal to $\mathcal{F}$ (Figure 3.8). $\qquad\square$

To simplify the notations, we shall from now on parametrize a vanishing cycle by S^1 rather than by the interval $[0, 1]$; an extension of a vanishing cycle $\alpha_0 \colon S^1 \to L_0$ is parametrized as $(\alpha_t \colon S^1 \to L_t)_{0 \le t < \epsilon}$. Recall that D denotes the unit disk in $\mathbb{C}$, so S^1 is the boundary of D.

The existence of the compact leaf will now follow by the next two lemmas.

Lemma 3.17 *Let α_0 be a positive simple vanishing cycle. Then there exist a positive extension $(\alpha_t)_{0 \le t < \epsilon}$ of α_0 and an immersion*

$$A \colon D \times (0, \epsilon) \longrightarrow M$$

such that $A_t = A(\,\text{-}\,, t) \colon D \to L_t$ extends $\alpha_t \colon S^1 \to L_t$ and lifts to an embedding $\tilde{A}_t \colon D \to \tilde{L}_t$ into the universal covering space $\tilde{L}_t$ of the leaf L_t, for any $0 < t < \epsilon$.

REMARK. We will write $D_t = A_t(D)$ and $\tilde{D}_t = \tilde{A}_t(D)$. As before we shall write $\tilde{\alpha}_t$ for the lift of α_t to the universal cover $\tilde{L}_t$ of L_t. Note

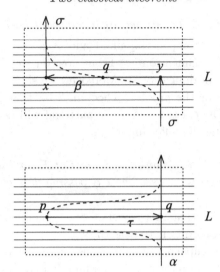

Fig. 3.8. Modifications of $\beta \cup \sigma$ and of $\alpha \cup \tau \cup \tau^{-1}$

that each $\tilde{L}_t$ is a non-compact connected orientable surface, and hence a copy of $\mathbb{R}^2$ (see Hirsch (1976), Exercise 3, page 207). Indeed, if L_t is compact, non-compactness of $\tilde{L}_t$ follows from Corollary 3.14.

Proof (of Lemma 3.17) For any (positive) extension (α_t) for which $\tilde{\alpha}_t$ is simple $(t > 0)$ we can extend $\tilde{\alpha}_t$ to an embedding $\tilde{A}_t \colon D \to \tilde{L}_t$. Then we project $\tilde{A}_t$ down to L_t to obtain A_t. We should now do this in such a way that (A_t) will be a smooth family.

First choose a normal vector field X to $\mathcal{F}$ and let T_x be the integral curve of X with $x \in T_x$, for any $x \in M$. In particular, T_x is a transversal section of $\mathcal{F}$, which we shall refer to as the normal transversal section through x. Let $(\alpha_t)_{0 \le t < \epsilon'}$ be a positive extension of α_0 obtained by holonomy with respect to the normal transversal sections.

Since α_0 is simple, there exists $0 < \epsilon < \epsilon'$ such that each $\tilde{\alpha}_t$ is simple, $0 < t \le \epsilon$. Now extend $\tilde{\alpha}_\epsilon$ to an embedding $\tilde{A}_\epsilon \colon D \to \tilde{L}_\epsilon$ and let A_ϵ be the composition of $\tilde{A}_\epsilon$ with the projection $\tilde{L}_\epsilon \to L_\epsilon$.

Using holonomy with respect to the normal transversal sections, we can extend A_ϵ to a smooth immersion $A \colon D \times (\delta, \epsilon] \to M$, for some $0 \le \delta < \epsilon$, such that $A_t = A(\,\text{-}\,, t) \colon D \to L_t$ extends α_t and lifts to $\tilde{L}_t$ as an embedding, for $\delta < t \le \epsilon$. Take the smallest $\delta \ge 0$ such that such an A can be defined. We claim that $\delta = 0$.

Assume that $\delta > 0$. Then $\tilde{\alpha}_\delta$ is a simple curve, so we can define an embedding $\tilde{B}_\delta\colon D \to \tilde{L}_\delta$ extending $\tilde{\alpha}_\delta$ and $B_\delta\colon D \to L_\delta$ as its projection down to L_δ, as before. Again, we can use holonomy with respect to normal transversal sections to obtain a smooth immersion

$$B\colon D \times (\delta - \mu, \delta + \mu) \longrightarrow M \, ,$$

for some small $\mu > 0$, such that $B_t = B(\,\cdot\,,t)\colon D \to L_t$ extends α_t and lifts to $\tilde{L}_t$ as an embedding, for $\delta - \mu < t < \delta + \mu$.

Take an $r \in (\delta, \delta + \mu)$. Since $\tilde{A}_r$ and $\tilde{B}_r$ are two embeddings of D into $\tilde{L}_r \cong \mathbb{R}^2$ which coincide on the boundary (they are both extending $\tilde{\alpha}_r$), there exists a diffeomorphism $f\colon D \to D$ with $f|_{S^1} = \mathrm{id}$ such that $\tilde{A}_r = \tilde{B}_r \circ f$. In particular, $A_r = B_r \circ f$. But both A and B were obtained by holonomy with respect to the same normal transversal sections, and hence it follows that $A_t = B_t \circ f$ for any $t \in (\delta, \delta + \mu)$. Now we can smoothly extend A to $D \times (\delta - \mu, \epsilon]$ by $A_t = B_t \circ f$, $t \in (\delta - \mu, \delta + \mu)$. Note that the extended A satisfies all the required properties. This shows that the smallest δ as above is in fact 0. $\qquad\square$

Lemma 3.18 *Under the same conditions and with the same notations as in Lemma 3.17, there exist a leaf $L \neq L_0$ and a decreasing sequence (t_m), converging to 0, such that $L_{t_m} = L$ and $D_{t_m} \subset \mathrm{Int}(D_{t_{m+1}})$ for any $m = 1, 2, \dots$. Moreover, $\tilde{D}_{t_m} \subset \tilde{D}_{t_{m+1}}$ and $\bigcup_m \tilde{D}_{t_m} = \tilde{L}$.*

Proof Since α_0 is not contractible in L_0, the map A cannot be extended continuously with one more disk $A_0\colon D \to L_0$ (to get an extension $D \times [0, \epsilon) \to M$). Thus, there is an $x \in \mathrm{Int}(D)$ such that the segment $T = \{A(x,t) \,|\, 0 < t < \epsilon\}$ does not have a unique limit point as $t \to 0$. On the other hand, by compactness of M, there is some limit point $r \in M$ of T as $t \to 0$, which lies on a leaf L. Choose a foliation chart U around r. There are points $r_m = A(x, s_m) \in U$, $m = 1, 2, \dots$, on the segment T which converge to r. Since T is transversal, we can clearly choose these points to be on L, and also we can assume that $L \neq L_0$ (Figure 3.9).

Write $D_m = D_{s_m}$. (This notation is unambiguous if we take $\epsilon < 1$.) First note that $r \in \mathrm{Int}(D_m)$ for large m. For if not, then since (r_m) converges to r and $r_m \in \mathrm{Int}(D_m)$, there must be points $q_m \in \partial D_m = \alpha_{s_m}(S^1)$ which converge to r. But the ∂D_m converge to $\alpha_0(S^1) \subset L_0$, so that $r \in L_0$, a contradiction.

Next, we claim that we can assume that the curves ∂D_m are disjoint. Indeed, start with ∂D_1. If each ∂D_m meets ∂D_1, then the sequence (∂D_m) has a limit point on ∂D_1, contradicting the fact that $\partial D_1 \subset L$

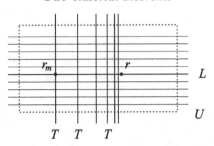

Fig. 3.9. Choice of the points r_m

while (∂D_m) converges to L_0. Thus there exists $m_2 > 0$ such that $\partial D_{m_2} \cap \partial D_1 = \emptyset$. Proceeding in this way, we find the desired subsequence (m_k) so that ∂D_{m_k} are disjoint. Thus, we will assume that ∂D_m are disjoint.

Consider the universal cover $\tilde{L}$ of L, and choose a base-point $\tilde{r} \in \tilde{L}$ above r. Since $r \in \text{Int}(D_m)$, we can lift the disks D_m to disks $\tilde{D}_m$ in $\tilde{L}$ such that $\tilde{r} \in \text{Int}(\tilde{D}_m)$. Then $\partial \tilde{D}_m$ is a lift of ∂D_m. Since α_0 is a simple vanishing cycle, $\partial \tilde{D}_m$ has no self-intersections, i.e. it is a simple curve in $\tilde{L} \cong \mathbb{R}^2$. Since the $\partial \tilde{D}_m$ are disjoint, they bound regions which are either disjoint or strictly contained in each other. The latter must be the case since they have a common point $\tilde{r}$. Since (∂D_m) has no limit points in L, a similar argument as before shows that we can assume that $\tilde{D}_m \subset \text{Int}(\tilde{D}_{m+1})$, after replacing (D_m) with a subsequence if necessary, and that $\bigcup_m \tilde{D}_m = \tilde{L}$. Since the quotient map $\tilde{L} \to L$ is open, we have $D_m \subset \text{Int}(D_{m+1})$ as well. $\qquad \square$

Theorem 3.19 *If α_0 is a positive simple vanishing cycle in L_0, then the leaf L_0 is compact.*

Proof Suppose that L_0 is not compact. Again we shall use the notations from Lemma 3.17 and Lemma 3.18. In particular, $(\alpha_t)_{0 \le t < \epsilon}$ denotes the positive extension of α_0. We may assume that this extension has the property that if $\alpha_t(z) = \alpha_{t'}(z')$ for some $0 \le t, t' < \epsilon$ then $\alpha_0(z) = \alpha_0(z')$ (see Remark 3.12 (1)). Furthermore, we have the immersion A as in Lemma 3.17. Write $p = \alpha_0(1)$, where we choose α_0 to be parametrized in such a way that p is not a point of self-intersection of α_0. By Lemma 3.16, there is a closed transversal curve γ through p. By modifying γ and by shrinking ϵ, we can assume that

(a) γ is positively oriented, $\gamma \colon [0, 1] \to M$, $\gamma(0) = \gamma(1) = p$,

(b) $\gamma(t) = \alpha_t(1)$ for $0 \le t < \epsilon$, and

(c) if $\gamma(t) = \alpha_{t'}(z)$ then $z = 1$ and $t = t'$, for any $t \in [0,1]$, $z \in S^1$ and $0 \le t' < \epsilon$.

(Modify γ to coincide with $\alpha(1, \text{-})$ on $[0, \epsilon)$ and make it transversal to $\alpha(S^1, [0, \epsilon))$ outside $[0, \epsilon)$. Then $\gamma|_{[\epsilon,1)}$ hits $\alpha(S^1, [0, \epsilon))$ in finitely many points, which we can get rid of by shrinking ϵ. Furthermore, $\gamma|_{[0,\epsilon)}$ hits $\alpha(S^1, [0, \epsilon))$ only in $\{\alpha_t(1) \,|\, 0 \le t < \epsilon\}$, by the property of Remark 3.12 (1) just mentioned and the fact that p is a simple point of α_0.)

Choose $m > 0$ and consider the restriction of A

$$A_m \colon D \times [t_{m+1}, t_m] \longrightarrow M .$$

The map A_m is a proper immersion. By Lemma 3.18, A_m factors as a proper immersion $f \colon Y \to M$ through the space Y, obtained as a quotient of $D \times [t_{m+1}, t_m]$ after identifying $D \times \{t_m\}$ with a suitable subspace of $D \times \{t_{m+1}\}$ (Figure 3.10). In particular, f has finite fibres. The curve

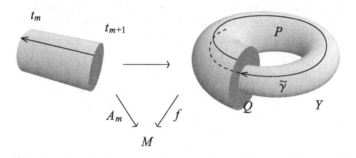

Fig. 3.10. The space Y

$\gamma|_{[t_{m+1}, t_m]}$ can be lifted to a (non-closed) curve $\tilde{\gamma}$ in Y. This lift can be extended to a lift $\tilde{\gamma} \colon [t_{m+1}, t_m + \delta] \to Y$ of the curve $\gamma|_{[t_{m+1}, t_m+\delta]}$ such that $\tilde{\gamma}((t_m, t_m + \delta)) \subset \text{Int}(Y)$. In fact, since f is a proper immersion, this lift is unique and can be prolonged if $\tilde{\gamma}(t_m + \delta) \in \text{Int}(Y)$.

Now the boundary of Y consists of two parts: one corresponds to $S^1 \times [t_{m+1}, t_m]$ which we denote by P, and the rest will be denoted by Q. Observe now that $\tilde{\gamma}(t_m + \delta)$ cannot lie in part Q because of the orientation of γ, but also it cannot lie in P, by assumption (c) on γ. This shows that $\tilde{\gamma}$ will never hit the boundary of Y, so it can be extended to a curve $\tilde{\gamma} \colon [t_{m+1}, \infty) \to Y$ such that $f(\tilde{\gamma}(t + k)) = \gamma(t)$ for any $t \in [0,1]$ and $k \in \mathbb{N}$ with $k + t \ge t_{m+1}$. Since $\tilde{\gamma}$ is clearly not periodic, this contradicts the fact that f has finite fibres. $\qquad\square$

Proposition 3.20 *If $H^1_{\mathrm{dR}}(M) = 0$ then any compact leaf of $\mathcal{F}$ is a torus.*

Proof Let L be a compact leaf of $\mathcal{F}$. It follows that L is the compact oriented surface T_g of some genus $g \geq 0$: we have to prove that $g = 1$, or equivalently, that the Euler characteristic $\chi(L) = 2 - 2g$ is zero.

To begin with, consider a small tubular neighbourhood N of L. This is a line bundle over L, which is oriented since the foliation is transversely oriented. Thus $N - L$ has two components, a positive one N^+, and a negative one N^-. Now the Mayer–Vietoris sequence for the open cover $M = (M - L) \cup N$ starts like

$$0 \longrightarrow H^0_{\mathrm{dR}}(M) \longrightarrow H^0_{\mathrm{dR}}(M - L) \oplus H^0_{\mathrm{dR}}(N) \longrightarrow H^0_{\mathrm{dR}}(N - L) \longrightarrow 0 \,.$$

Since N is connected and $N - L$ has two components, we see that $M - L$ has two components, say $V = V^+$ and V^-, such that $N^+ \subset V^+$ and $N^- \subset V^-$. In particular, $\bar{V}$ is a foliated manifold with boundary equal to the leaf L; since foliation is transversely orientable, there exists a nowhere vanishing normal field X on $\bar{V}$ which is transversal to the boundary L of $\bar{V}$.

Now consider $L \cong T_g$ embedded in $S^3 = \mathbb{R}^3 \cup \{\infty\}$, so that $\infty \notin T_g$, and let C be the component of $S^3 - T_g$ with $\infty \in C$. Consider the space

$$G = \bar{C} \cup_{T_g} V \,.$$

Now we can extend the field X to a vector field $\bar{X}$ on G such that $\bar{X}$ has exactly $g + 1$ singularities: we add one at ∞ with index 1, which looks like the gradient of the function $x^2 + y^2 + z^2$ (at zero), and g singularities, one in each 'hole' of T_g, with index -1, which look like the gradient of the function $x^2 + y^2 - z^2$ (Figure 3.11). The sum of indices is thus $1 - g$, so the Euler characteristic of G is $1 - g$ by the Poincaré–Hopf theorem. On the other hand, the Euler characteristic of an odd dimensional (orientable) compact manifold is zero, by Poincaré duality. This yields $g = 1$. $\qquad\square$

Corollary 3.21 *If $\pi_1(M)$ is finite, then there exists a compact leaf of $\mathcal{F}$ which is a torus.*

REMARK. This torus is the boundary of a submanifold $V \subset M$ of dimension 3 which is a union of leaves of $\mathcal{F}$.

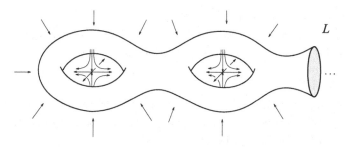

Fig. 3.11. The vector field $\bar{X}$ on G

3.2.3 Existence of a Reeb component

In this subsection M is a compact connected oriented manifold of dimension 3 with finite fundamental group, and $\mathcal{F}$ is a transversely oriented foliation of M of codimension 1.

Theorem 3.22 *The foliated manifold* $(M, \mathcal{F})$ *has a Reeb component.*

REMARK. The theorem asserts that there exists a topological embedding $D \times S^1 \to M$ which maps the leaves of a Reeb foliation on the solid torus $D \times S^1$ onto the leaves of $\mathcal{F}$.

Proof (of Theorem 3.22) By Theorem 3.13 and Proposition 3.15 there exists a simple (positive) vanishing cycle α_0 in a leaf L_0, and by Theorem 3.19 the leaf L_0 is compact. Further, Proposition 3.20 implies that L_0 is a torus.

First we will show that we can take α_0 to be a simple curve. Let v and w be the standard loops in $L_0 \cong T^2$ which generate the fundamental group of L_0. Thus there exist uniquely determined $p, q \in \mathbb{Z}$ such that α_0 is in the same homotopy class as $v^p w^q$.

Observe that if δ_0 is a loop in L_0 such that δ_0^k is a positive vanishing cycle, then δ_0 is a positive vanishing cycle as well. Indeed, first note that δ_0 has trivial positive holonomy. Let (δ_t) be a positive extension of δ_0 obtained using holonomy. Since δ_0^k is a vanishing cycle, the loop δ_t^k is contractible in its leaf. But we know that all the leaves of $\mathcal{F}$ have $\mathbb{R}^2$ for their universal covering space; in particular, their fundamental groups are torsion free, as no finite group can act freely on $\mathbb{R}^2$. (Here is one way to see this. If a finite group G acts freely on $\mathbb{R}^2$, then so does any of its cyclic subgroups, thus we may assume that G is cyclic. Since $\mathbb{R}^2$ is contractible, the two-dimensional manifold $X = \mathbb{R}^2/G$ is a

$K(G, 1)$ space. In particular, $H_0(X, \mathbb{Z}) = \mathbb{Z}$, $H_1(X, \mathbb{Z}) = G$, $H_i(X, \mathbb{Z}) = H_i(G, \mathbb{Z})$. But $H_i(X, \mathbb{Z}) = 0$ if $i > 2$, for dimensional reasons, while $H_i(G, \mathbb{Z})$ is periodic with period 2 (Mac Lane (1963)). This gives a contradiction.) Thus δ_t is itself contractible in its leaf, hence δ_0 is a vanishing cycle.

This shows that we can assume without loss of generality that p and q are relatively prime. But then α_0 is a generator of $\pi_1(L_0)$ (there exist $a, b \in \mathbb{Z}$ with $ap - bq = 1$, hence the matrix

$$\begin{pmatrix} p & b \\ q & a \end{pmatrix}$$

is invertible over $\mathbb{Z}$), and we can choose the diffeomorphism $g \colon T^2 \to L_0$ in such a way that $\alpha_0 = v$, i.e. $\alpha_0(z) = g(z, 1)$ for $z \in S^1$. In particular, it is a simple curve and hence a simple vanishing cycle. Put $\beta_0 = g(1, \text{-})$.

In Lemma 3.17 we associated to the vanishing cycle α_0 the immersion $A \colon D \times (0, \epsilon) \to M$. This was done using holonomy with respect to the normal transversal sections, which are the integral curves of a fixed normal vector field X on M.

Note that α_0 has of course trivial positive holonomy. On the other hand, β_0 must have non-trivial positive holonomy, otherwise the leaves L_t would be tori for small t by the local Reeb stability theorem. Hence the positive holonomy group of L_0 is $\mathbb{Z}$, and we can assume that $\mathrm{hol}(\beta_0)$ shrinks the positive part of a transversal section.

Now write $\beta_z = g(z, \text{-})$ for $z \in S^1$ (thus $\beta_1 = \beta_0$). Each β_z is a closed curve with the 'same' holonomy as β_0. By holonomy of those curves, using the chosen normal transversal sections, we obtain a smooth map

$$Z \colon W = S^1 \times [0, 1] \times [0, \delta] \longrightarrow M$$

for some small $0 < \delta < \epsilon$. More precisely, $Z(z, h, t)$ is the transport along β_z during time h starting at $\alpha_t(z)$. Thus Z satisfies the following conditions:

  (i) $Z(z, h, 0) = \beta_z(e^{2\pi i h})$,
  (ii) $Z(z, 0, t) = \alpha_t(z)$,
  (iii) $Z(z, h, t) \in L_t$, and
  (iv) $Z(z, h, [0, \delta])$ is a normal section of $\mathcal{F}$.

Of course we choose δ so small that the transversal sections of (iv) are disjoint for different $(z, h) \in S^1 \times [0, 1)$. Further, for any $z \in S^1$ and $t \in [0, \delta]$ we have $Z(z, 1, t) = Z(z, 0, \zeta_z(t))$ for some $\zeta_z(t) \in [0, \delta]$; in other words, ζ_z is a reparametrization of the holonomy of β_z. But

$\zeta_z(t)$ is clearly a locally constant function of z, therefore constant, and we shall write $\zeta = \zeta_z$. In other words, there is a unique smooth map $\zeta \colon [0, \delta] \to [0, \delta]$ with $Z(z, 1, t) = Z(z, 0, \zeta(t))$ for any $z \in S^1$ and $t \in [0, \delta]$.

Let W' be the quotient of W obtained after identifying $(z, 1, t)$ with $(z, 0, \zeta(t))$ for any $z \in S^1$ and $t \in [0, \delta]$. Then it is easy to see that Z factors as an embedding $Z' \colon W' \to M$. Note also that we have $\zeta(0) = 0$ and $\zeta(t) < t$ for any $t > 0$: for if $\zeta(t) = t$ for some t then L_t is a torus and α_t is not contractible, a contradiction (Figure 3.12). For any

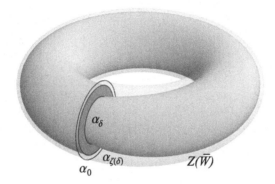

Fig. 3.12. The image of Z

$0 < t \leq \delta$, α_t is a simple closed curve in L_t and clearly divides L_t into two components, one of them being diffeomorphic to $\mathbb{R}^2 - D$. From the fact that α_t is contractible in L_t and that L_t has $\mathbb{R}^2$ for its universal covering space it follows that $L_t \cong \mathbb{R}^2$. In particular, D_t is an embedded disk in L_t. Furthermore, $L_t = L_{t'}$ for $0 < t' < t < \delta$ if and only if $t' = \zeta^k(t)$ for some $k \in \mathbb{N}$.

In Lemma 3.17 we have shown that for δ small enough (α_t) can be extended to an immersion $A \colon D \times (0, \delta] \to M$, again by using holonomy with respect to the normal transversal sections. Adopting the notations of Subsection 3.2.2, we have $D_t \subset \mathrm{Int}(D_{\zeta(t)})$ and

$$D_t \cup \bigcup_{k=0}^{\infty} Z(S^1, [0, 1], \zeta^k(t)) = L_t$$

for any $0 < t \leq \delta$. As in the proof of Theorem 3.19, the restriction of A

$$A' \colon D \times [\zeta(\delta), \delta] \to M$$

factors as a map f through the space Y obtained as a quotient of the product $D \times [\zeta(\delta), \delta]$ (see Figure 3.10, replacing t_m with δ and t_{m+1} with $\zeta(\delta)$). We know from the proof of Theorem 3.19 that f is a proper immersion, but by our observations above it is clear that f is in fact an embedding. Now there is an obvious way to glue Y and W' together, obtaining a solid torus with a Reeb foliation, and combine Z' and f to an embedding of this torus into M which preserves the leaves. $\square$

4

Molino's theory

In Chapter 2 we introduced the notion of a Riemannian foliation: this is a foliation whose normal bundle is equipped with a metric which is, in an appropriate sense, invariant under transport along the leaves of the foliation. In this chapter we study some special classes of Riemannian foliations, and some ways of constructing them, with the ultimate goal of proving Molino's 'structure theorem'.

The most important class of Riemannian foliations in this chapter is that of transversely parallelizable ones. While an ordinary manifold is called parallelizable if its tangent bundle is trivial, a foliated manifold is called transversely parallelizable if its normal bundle is trivial, and if moreover a trivialization exists which is invariant under transport along the leaves. Intuitively speaking, these are the foliated manifolds whose 'space of leaves' is parallelizable.

A special class of transversely parallelizable foliations (to be discussed in Subsection 4.3.1 below) are the so-called Lie foliations. These are foliations defined as the kernel of a 'Maurer–Cartan' differential 1-form with values in a Lie algebra.

Another way of obtaining transversely parallelizable foliations, to be discussed in Subsection 4.2.2, is by pulling back a given Riemannian foliation on a manifold M to a suitable transverse frame bundle over M. This construction will form an important ingredient for Molino's structure theorem.

Any transversely parallelizable foliation on a compact connected manifold is homogeneous, in the sense that the group of global diffeomorphisms preserving the foliation acts transitively on the manifold (see Theorem 4.8. We begin this chapter with the discussion of such homogeneous foliations, and prove in particular that for any homogeneous foliation $\mathcal{F}$ of a manifold M one can construct another foliation $\mathcal{F}_{\mathrm{bas}}$

on M by thickening the leaves of $\mathcal{F}$ in a particular way. This thicker foliation is always strictly simple, and is uniquely characterized by the property that a smooth function on M is constant along the leaves of $\mathcal{F}_{\text{bas}}$ if and only if it is constant along the leaves $\mathcal{F}$ (see Theorem 4.3).

Our study of homogeneity and transverse parallelizability for foliations will naturally lead to Molino's theorem, already alluded to above. This theorem states that every Riemannian foliation on a compact connected manifold can be covered in a canonical way by a transversely parallelizable foliation. Moreover, on each of the associated thick leaves, this latter foliation restricts to a Lie foliation, whose Maurer–Cartan form takes values in a Lie algebra naturally associated to the original foliation.

4.1 Transverse parallelizability

In this section, we discuss the notions of homogeneity and transverse parallelizability for foliations, related by the fact that any transversely parallelizable foliation is homogeneous. Any transversely parallelizable foliation is Riemannian. We also discuss how the leaves of a homogeneous foliation can be thickened so as to give us an associated 'basic' foliation, again homogeneous, which is strictly simple.

4.1.1 Homogeneous foliations

Recall that a manifold M is said to be *homogeneous* if for any two points $x, y \in M$ there exists a diffeomorphism $M \to M$ which maps x to y. It is a standard fact that any second-countable Hausdorff manifold is homogeneous. However, for non-Hausdorff manifolds the situation is more subtle; see Exercise 4.2 (3) below.

An *automorphism* of a foliated manifold $(M, \mathcal{F})$ is a diffeomorphism $\phi : M \to M$ which preserves the foliation, i.e. for which the image of any leaf is a leaf (Section 1.1), or equivalently, $d\phi(T(\mathcal{F})) = T(\mathcal{F})$. The automorphisms of $(M, \mathcal{F})$ form a group which we shall denote by $\text{Aut}(M, \mathcal{F})$. A foliation $\mathcal{F}$ of a manifold M is called *homogeneous* if the group $\text{Aut}(M, \mathcal{F})$ acts transitively on M, or in other words, if for any two points $x, y \in M$ there exists a diffeomorphism $\phi : M \to M$ which preserves the foliation and maps x to y. It is easy to see that this is always true if x and y lie on the same leaf (Exercise 4.2 (2) below), but if this is true for any two points there are several quite exclusive properties of the foliation that one can deduce. In particular, all the leaves of a homogeneous foliation are diffeomorphic. Among the examples of foliations we have mentioned so far in this book, the Kronecker foliation of the torus is one of the few non-trivial examples which is homogeneous. One of the special features of a homogeneous foliation is that it gives rise to another foliation, which is given by the fibres of a fibre bundle if the manifold is compact. Its description uses the concept of a basic function, which we discuss first.

Let $(M, \mathcal{F})$ be a foliated manifold. The vector fields on M which are tangent to the leaves of $\mathcal{F}$ form a Lie subalgebra of $\mathfrak{X}(M)$, which we denote by $\mathfrak{X}(\mathcal{F})$ (see Section 1.2). In other words, $\mathfrak{X}(\mathcal{F})$ consists of the sections of the tangent bundle $T(\mathcal{F})$ of the foliation $\mathcal{F}$. A smooth function f on M is called *basic* if it is constant along the leaves. Equivalently, a function f is basic if $X(f) = 0$ whenever $X \in \mathfrak{X}(\mathcal{F})$, briefly $\mathfrak{X}(\mathcal{F})(f) = 0$. The basic functions on $(M, \mathcal{F})$ form a subalgebra $\Omega^0_{\mathrm{bas}}(M, \mathcal{F})$ of $C^\infty(M)$. If a function f is defined locally on an open subset U of M, then f is called basic if it is basic with respect to the foliation $\mathcal{F}|_U$. Locally on a foliation chart we have a lot of basic functions; however, it may be impossible to extend them to M. Note that if X is a vector field on M such that $X(f) = 0$ for any locally defined basic function, then $X \in \mathfrak{X}(\mathcal{F})$.

Examples 4.1 (1) Let $f \colon M \to Q$ be a surjective submersion with connected fibres, and let $\mathcal{F}$ be the strictly simple foliation of M given by the fibres of f. Then we have $\Omega^0_{\mathrm{bas}}(M, \mathcal{F}) = C^\infty(Q)$.

(2) Let $(M, \mathcal{F})$ be a foliated manifold. We may choose a Haefliger cocycle $(s_i \colon U_i \to \mathbb{R}^q)$ which defines $\mathcal{F}$, and we may also assume that the submersions s_i have connected fibres. Then a function f on M is basic if and only if $f|_{U_i}$ is a pull-back along s_i of a function on $s_i(U_i)$, for any i.

(3) Let $\mathcal{F}$ be the Kronecker foliation of the torus T^2. Since any leaf of $\mathcal{F}$ is dense in T^2, any basic function is constant, so $\Omega^0_{\mathrm{bas}}(T^2, \mathcal{F}) = \mathbb{R}$.

Exercises 4.2 (1) Let M be a (second-countable Hausdorff) manifold and let $x, y \in M$. Show that there exists a diffeomorphism $M \to M$ which is the identity outside a compact subset of M and maps x to y.

(2) Let $(M, \mathcal{F})$ be a foliated manifold and let x and y be two points of M lying on the same leaf of $\mathcal{F}$. Show that there exists an automorphism of $(M, \mathcal{F})$ which maps x to y. The automorphism may be chosen such that it is the identity outside a compact set.

(3) Observe that a non-Hausdorff manifold need not be homogeneous. Show that in fact no second-countable non-Hausdorff manifold is ho-

mogeneous. (Hint: Suppose that M is a non-Hausdorff homogeneous manifold of dimension $n \geq 1$. By non-Hausdorffness and homogeneity of M any point x of M has the property that there exists another point of M which cannot be separated from x. If M is second-countable, there exists a countable cover $(U_i)_{i=1}^{\infty}$ of M open non-empty Hausdorff subsets of M. For any point $x \in U_1$ choose $i(x) \in \mathbb{N}$ such that a point which cannot be separated from x lies in $U_{i(x)}$. This implies that $V_{i(x)} = U_1 \cap U_{i(x)}$ is a non-empty open subset of U_1 which has x in its boundary. Now if we take $J = \{i(x) \mid x \in U_1\} \subset \mathbb{N}$, and if W_j is the boundary of V_j inside U_1, the family $(W_j)_{j \in J}$ is a countable cover of U_1 made up of closed subsets with empty interior, which is impossible.)

Let $\mathcal{F}$ be a homogeneous foliation of a manifold M. Let $\mathfrak{X}_{\mathrm{bas}}(\mathcal{F})$ be the vector space of those vector fields on M which vanish on the global basic functions,

$$\mathfrak{X}_{\mathrm{bas}}(\mathcal{F}) = \{X \in \mathfrak{X}(M) \mid X(\Omega^0_{\mathrm{bas}}(M, \mathcal{F})) = 0\} \ .$$

Clearly we have $\mathfrak{X}(\mathcal{F}) \subset \mathfrak{X}_{\mathrm{bas}}(\mathcal{F})$, while $\mathfrak{X}_{\mathrm{bas}}(\mathcal{F})$ is a Lie subalgebra of $\mathfrak{X}(M)$ and a module over $C^{\infty}(M)$. For any $x \in M$ put

$$E_x = \{X_x \mid X \in \mathfrak{X}_{\mathrm{bas}}(\mathcal{F})\} \subset T_x(M) \ .$$

Since $\mathcal{F}$ is homogeneous, the dimension of E_x is constant with respect to $x \in M$. Indeed, for any $\phi \in \mathrm{Aut}(M, \mathcal{F})$ we have $E_{\phi(x)} = (d\phi)_x(E_x)$. To see this, observe first that the composition of ϕ with a basic function is again basic. Hence if $X \in \mathfrak{X}_{\mathrm{bas}}(\mathcal{F})$ we have $d\phi(X)(f) = X(f \circ \phi) = 0$ and therefore $(d\phi)_x(X_x) \in E_{\phi(x)}$. It follows that $E = \bigcup_{x \in M} E_x$ is an involutive subbundle of $T(M)$, so it defines a foliation $\mathcal{F}_{\mathrm{bas}}$ on M with $T(\mathcal{F}_{\mathrm{bas}}) = E$, for which it holds that

$$\mathfrak{X}(\mathcal{F}_{\mathrm{bas}}) = \mathfrak{X}_{\mathrm{bas}}(\mathcal{F}) \ .$$

The foliation $\mathcal{F}_{\mathrm{bas}}$ will also be referred to as the *basic* foliation associated to $\mathcal{F}$. Since $\mathfrak{X}(\mathcal{F}) \subset \mathfrak{X}(\mathcal{F}_{\mathrm{bas}})$, it follows that any leaf of $\mathcal{F}$ is contained in a leaf of $\mathcal{F}_{\mathrm{bas}}$. Write $q = \mathrm{codim}\,\mathcal{F}$ and $q' = \mathrm{codim}\,\mathcal{F}_{\mathrm{bas}} \leq q$. So if L is a leaf of $\mathcal{F}_{\mathrm{bas}}$, the foliation $\mathcal{F}$ can be restricted to L, and $(L, \mathcal{F}|_L)$ is a foliated manifold of codimension $q' - q$.

Theorem 4.3 *Let $\mathcal{F}$ be a homogeneous foliation of M, and let $\mathcal{F}_{\mathrm{bas}}$ be the associated basic foliation. Then*

(i) $\Omega^0_{\mathrm{bas}}(M, \mathcal{F}) = \Omega^0_{\mathrm{bas}}(M, \mathcal{F}_{\mathrm{bas}})$, i.e. $\mathcal{F}$ and $\mathcal{F}_{\mathrm{bas}}$ have the same basic functions,

(ii) $\mathrm{Aut}(M, \mathcal{F}) \subset \mathrm{Aut}(M, \mathcal{F}_{\mathrm{bas}})$, *and* $\mathcal{F}_{\mathrm{bas}}$ *is again homogeneous,*

(iii) $\mathcal{F}_{\mathrm{bas}}$ *is strictly simple, i.e. the space of basic leaves* $M/\mathcal{F}_{\mathrm{bas}}$ *has the structure of a (Hausdorff) manifold such that the quotient projection* $\pi_{\mathrm{bas}} \colon M \to M/\mathcal{F}_{\mathrm{bas}}$ *is a submersion,*

(iv) *the projection* π_{bas} *induces an isomorphism between* $\Omega^0_{\mathrm{bas}}(M, \mathcal{F})$ *and* $C^\infty(M/\mathcal{F}_{\mathrm{bas}})$,

(v) *if the leaves of* $\mathcal{F}_{\mathrm{bas}}$ *are compact (e.g. if* M *is compact), then* $\pi_{\mathrm{bas}} \colon M \to M/\mathcal{F}_{\mathrm{bas}}$ *is a fibre bundle, and*

(vi) *if* $\mathcal{F}$ *is simple, then it is strictly simple and equal to* $\mathcal{F}_{\mathrm{bas}}$.

Proof (i) Since $\mathfrak{X}(\mathcal{F}) \subset \mathfrak{X}(\mathcal{F}_{\mathrm{bas}})$ we have $\Omega^0_{\mathrm{bas}}(M, \mathcal{F}_{\mathrm{bas}}) \subset \Omega^0_{\mathrm{bas}}(M, \mathcal{F})$. On the other hand, if $f \in \Omega^0_{\mathrm{bas}}(M, \mathcal{F})$ then $\mathfrak{X}(\mathcal{F}_{\mathrm{bas}})(f) = 0$ by definition of $\mathcal{F}_{\mathrm{bas}}$, therefore $f \in \Omega^0_{\mathrm{bas}}(M, \mathcal{F}_{\mathrm{bas}})$.

(ii) We showed already that $E_{\phi(x)} = (d\phi)_x(E_x)$ for any $\phi \in \mathrm{Aut}(M, \mathcal{F})$ and $x \in M$, so $d\phi(T(\mathcal{F}_{\mathrm{bas}})) = T(\mathcal{F}_{\mathrm{bas}})$.

(iii) We first remark that for any point $x \in M$,

$$E_x = \{\xi \in T_x M \,|\, df(\xi) = 0 \text{ for any basic function } f\}\,.$$

To see this, observe that the right hand side itself defines a subbundle of $T(M)$ because $\mathcal{F}$ is homogeneous, and hence any tangent vector ξ from the right hand side can be extended to a section of this subbundle, i.e. to a vector field in $\mathfrak{X}(\mathcal{F}_{\mathrm{bas}})$. Next, note that we can choose basic functions $f_1, \ldots, f_{q'}$ such that $(df_1)_x, \ldots, (df_{q'})_x$ are linearly independent and $E_x = \bigcap_{i=1}^{q'} \mathrm{Ker}(df_i)_x$. The same is then true at any point in a small neighbourhood U of x, and hence $\mathcal{F}_{\mathrm{bas}}|_U$ is given by the fibres of the map $s = (f_1, \ldots, f_{q'}) \colon M \to \mathbb{R}^{q'}$ restricted to U. We may shrink U if necessary so that the fibres of $s|_U$ are connected. Now since the functions $f_1, \ldots, f_{q'}$ are globally defined and constant along the leaves, any leaf of $\mathcal{F}_{\mathrm{bas}}$ intersects U in at most one plaque. Note that this implies that any leaf is closed and has trivial holonomy. Furthermore, the map s induces a homeomorphism of $\pi_{\mathrm{bas}}(U)$ with $s(U) \subset \mathbb{R}^{q'}$, which gives a local chart for a smooth structure on $M/\mathcal{F}_{\mathrm{bas}}$. The change-of-charts homeomorphisms obtained in this way are given by the holonomy transport, hence we obtain a smooth structure on $M/\mathcal{F}_{\mathrm{bas}}$.

Finally, we need to prove that $M/\mathcal{F}_{\mathrm{bas}}$ is Hausdorff. It is a homogeneous manifold precisely because $(M, \mathcal{F}_{\mathrm{bas}})$ is a homogeneous foliation. Since $M/\mathcal{F}_{\mathrm{bas}}$ is second-countable as well, it is Hausdorff (see Exercise 4.2 (3)).

(iv) This is a consequence of (i) and (iii).

(v) This follows from (iii) by the local Reeb stability theorem (Theorem 2.9).

(vi) Suppose that $\mathcal{F}$ is given by the connected components of a submersion $\phi \colon M \to Q$. Put $q = \operatorname{codim} \mathcal{F} = \dim Q$, take any point $x \in M$ and choose a smooth function $f = (f_1, \ldots, f_q) \colon Q \to \mathbb{R}^q$ which is a diffeomorphism on an open neighbourhood V of $\phi(x)$ in Q. Now $f_1 \circ \phi, \ldots, f_q \circ \phi$ are basic functions on $(M, \mathcal{F})$ which separate leaves in $\phi^{-1}(V)$. This implies that $\dim \mathcal{F} = \dim \mathcal{F}_{\mathrm{bas}}$, hence $\mathcal{F} = \mathcal{F}_{\mathrm{bas}}$ and $\mathcal{F}$ is strictly simple by (iii). $\square$

4.1.2 Transversely parallelizable foliations

Let $\mathcal{F}$ be a foliation of codimension q on a manifold M of dimension n. In general, the Lie subalgebra $\mathfrak{X}(\mathcal{F})$ is not a Lie ideal in $\mathfrak{X}(M)$, but it is clearly a Lie ideal in the Lie subalgebra

$$L(M, \mathcal{F}) = \{ Y \in \mathfrak{X}(M) \, | \, [\mathfrak{X}(\mathcal{F}), Y] \subset \mathfrak{X}(\mathcal{F}) \} \,.$$

This is exactly the algebra of projectable vector fields (Remark 2.7 (7)). Note that $L(M, \mathcal{F})$ is a module over the algebra of basic functions. Indeed, if $Y \in L(M, \mathcal{F})$ and if f is basic, then for any $X \in \mathfrak{X}(\mathcal{F})$ we have $[X, fY] = X(f)Y + f[X, Y] \in \mathfrak{X}(\mathcal{F})$ since $X(f) = 0$ and Y is projectable. Next, the definition of the Lie bracket implies that the derivative of a basic function in the direction of a projectable vector field is again basic. In fact, the converse is also true: if Y is a vector field on M such that $Y(f)$ is a (locally defined) basic function for any locally defined basic function f, then Y is projectable.

We shall denote the quotient Lie algebra $L(M, \mathcal{F})/\mathfrak{X}(\mathcal{F})$ by $l(M, \mathcal{F})$,

$$0 \longrightarrow \mathfrak{X}(\mathcal{F}) \longrightarrow L(M, \mathcal{F}) \longrightarrow l(M, \mathcal{F}) \longrightarrow 0 \,.$$

This Lie algebra is also a module over $\Omega^0_{\mathrm{bas}}(M, \mathcal{F})$. For any $Y \in L(M, \mathcal{F})$ we shall write $\bar{Y}$ for the projection of Y in $l(M, \mathcal{F})$. Elements of $l(M, \mathcal{F})$ are called *transverse* vector fields on $(M, \mathcal{F})$, and act as derivations on $\Omega^0_{\mathrm{bas}}(M, \mathcal{F})$ by $\bar{Y}(f) = Y(f) + \mathfrak{X}(\mathcal{F})(f) = Y(f)$. Note that they can be viewed as certain sections of the normal bundle $N(\mathcal{F})$ of the foliation. Such a section $\sigma \colon M \to N(\mathcal{F})$ is a transverse vector field if and only if it can be locally projected along a submersion which locally defines the foliation, as the following example illustrates.

Examples 4.4 (1) Let $(x_1, \ldots, x_p, y_1, \ldots, y_q) \colon U \to \mathbb{R}^p \times \mathbb{R}^q$ be a surjective foliation chart for $\mathcal{F}$. Then the vector fields $\frac{\partial}{\partial x_1}, \ldots, \frac{\partial}{\partial x_p}$ generate the module $\mathfrak{X}(\mathcal{F})|_U$ over $C^\infty(M)|_U$. Take any vector field Y on M and write $Y|_U = \sum_i a_i \frac{\partial}{\partial x_i} + \sum_j b_j \frac{\partial}{\partial y_j}$. Then we have

$$\left[\frac{\partial}{\partial x_k}, Y\right] = \sum_i \frac{\partial a_i}{\partial x_k} \frac{\partial}{\partial x_i} + \sum_j \frac{\partial b_j}{\partial x_k} \frac{\partial}{\partial y_j} .$$

Thus Y is projectable if for any such local chart one has

$$\frac{\partial b_j}{\partial x_k} = 0 , \qquad j = 1, \ldots, q \quad k = 1, \ldots, p ,$$

or in other words, if the functions b_i are basic for $\mathcal{F}|_U$. Equivalently, this means that Y can be projected to the quotient $U/\mathcal{F} \cong \mathbb{R}^q$. In this case we have

$$\bar{Y}|_U = \sum_j b_j \frac{\partial}{\partial y_j} + \mathfrak{X}(\mathcal{F}) .$$

This is a section of the normal bundle of $\mathcal{F}|_U$ which can be projected to $\mathbb{R}^q$ since the functions b_j are constant along the leaves. We see that locally we have a lot of projectable vector fields, but globally this may not be the case.

(2) Let $f \colon M \to Q$ be a surjective submersion with connected fibres, and let $\mathcal{F}$ be the strictly simple foliation of M given by the fibres of f. Then $l(M, \mathcal{F}) = \mathfrak{X}(Q)$.

(3) Let $(M, \mathcal{F})$ be a foliated manifold, and choose a Haefliger cocycle $(s_i \colon U_i \to \mathbb{R}^q)$ which defines $\mathcal{F}$ such that the submersions s_i have connected fibres. Then ds_i induces an isomorphism $N_x(\mathcal{F}) \to \mathbb{R}^q$ in any point $x \in U_i$. A section σ of $N(\mathcal{F})$ is a transverse vector field if and only if $\sigma|_{U_i}$ is a pull-back of a vector field on $s_i(U_i)$ along s_i, for any i.

(4) For the Kronecker foliation $\mathcal{F}$ of the torus T^2 one has $l(T^2, \mathcal{F}) = \mathbb{R}$.

Lemma 4.5 *Let $\mathcal{F}$ be a homogeneous foliation of a manifold M. Then $L(M, \mathcal{F}) \subset L(M, \mathcal{F}_{\text{bas}})$, where $\mathcal{F}_{\text{bas}}$ is the basic foliation of M associated to $\mathcal{F}$.*

Proof Let $Y \in L(M, \mathcal{F})$ and take any $X \in \mathfrak{X}(\mathcal{F}_{\text{bas}})$. For any basic function f we have

$$[X, Y](f) = X(Y(f)) - Y(X(f)) = 0$$

since $Y(f)$ is also basic and X annihilates basic functions by definition. This implies that $[X, Y] \in \mathfrak{X}(\mathcal{F}_{\text{bas}})$, and hence $Y \in L(M, \mathcal{F}_{\text{bas}})$. $\qquad\square$

Projectable vector fields can also be characterized by the property that their flows preserve the foliation (note that the flow may be defined only locally if M is not compact). In order to see that this property is necessary (a similar computation shows that it is sufficient as well), let Y be a projectable vector field and let μ be the (locally defined) flow of Y. To see that each $\mu(t, \text{-}) = e^{tY}$ (locally) preserves the foliation it is enough to prove that the composition of a locally defined basic function f with e^{tY} is again basic. To this end, take any $X \in \mathfrak{X}(\mathcal{F})$. We have to show that $X(f \circ e^{tY}) = 0$ for any t. This is clearly true for $t = 0$, so it is sufficient to show that $\frac{\partial}{\partial t} X(f \circ e^{tY}) = 0$. But the derivations X and $\frac{\partial}{\partial t}$ commute, so the definition of the flow e^{tY} and the fact that $Y(f)$ is also basic give us

$$
\begin{aligned}
\frac{\partial}{\partial t} X(f \circ e^{tY}) &= X\left(\frac{\partial}{\partial t}(f \circ e^{tY})\right) \\
&= X\left(\frac{\partial e^{tY}}{\partial t}(f)\right) \\
&= X(Y(f)) \\
&= 0 .
\end{aligned}
$$

A foliated manifold $(M, \mathcal{F})$ is called *transversely parallelizable* if there exist transverse vector fields $\bar{Y}_1, \ldots, \bar{Y}_q \in l(M, \mathcal{F})$ which form a global frame for the normal bundle of the foliation $\mathcal{F}$. In that case the fields $\bar{Y}_1, \ldots, \bar{Y}_q$, also referred to as a *transverse parallelism* for $(M, \mathcal{F})$, form a basis of the module $l(M, \mathcal{F})$ over $\Omega^0_{\text{bas}}(M, \mathcal{F})$. Indeed, we can write any section σ of the normal bundle as $\sigma = a_1 \bar{Y}_1 + \cdots + a_q \bar{Y}_q$ for some smooth functions $a_1, \ldots, a_q \in C^\infty(M)$, and this section σ is a transverse vector field if and only if all the functions $a_1, \ldots, a_q$ are basic (exercise).

Examples 4.6 (1) Let $\omega_1, \ldots, \omega_q$ be closed, (pointwise) linearly independent 1-forms on a manifold M. They define a foliation $\mathcal{F}$ on M of codimension q by

$$
T(\mathcal{F}) = \bigcap_{i=1}^{q} \text{Ker}(\omega_i) .
$$

Let $Y_1, \ldots, Y_q$ be vector fields on M such that $\omega_i(Y_j) = \delta_{ij}$. These vector fields are projectable, since we have for any $X \in \mathfrak{X}(\mathcal{F})$

$$
\omega_i([X, Y_j]) = X(\omega_i(Y_j)) - Y_j(\omega_i(X)) - 2d\omega_i(X, Y_j) = X(\delta_{ij}) = 0 .
$$

Therefore $\bar{Y}_1, \ldots, \bar{Y}_q$ form a transverse parallelism for $(M, \mathcal{F})$.

(2) The trivial foliation of dimension 0 on a manifold M is transversely parallelizable if and only if the manifold M is parallelizable (i.e. the tangent bundle $T(M)$ is trivial).

Proposition 4.7 *If $(M, \mathcal{F})$ is transversely parallelizable, then all the leaves of $\mathcal{F}$ have trivial holonomy, and there exists a transverse metric on $(M, \mathcal{F})$, i.e. $(M, \mathcal{F})$ can be given the structure of a Riemannian foliation.*

Proof Recall that the image of a projectable vector field Z on $(M, \mathcal{F})$ under the quotient map $L(M, \mathcal{F}) \to l(M, \mathcal{F})$ is denoted by $\bar{Z}$. Let $\bar{Y}_1, \ldots, \bar{Y}_q$ be a transverse parallelism on $(M, \mathcal{F})$. Then define a Riemannian structure $\langle - , - \rangle$ on $N(\mathcal{F})$ by $\langle \bar{Y}_i, \bar{Y}_j \rangle = \delta_{ij}$. Now the transverse metric g on $T(M)$ (see Section 2.2) is given by

$$g(Y, Z) = \langle \bar{Y}, \bar{Z} \rangle .$$

The kernel of g_x is indeed $T_x(\mathcal{F})$, and by definition of $L_X g$ we have for any $X \in \mathfrak{X}(\mathcal{F})$

$$
\begin{aligned}
L_X g(Y_i, Y_j) &= X(\langle \bar{Y}_i, \bar{Y}_j \rangle) - \langle \overline{[X, Y_i]}, \bar{Y}_j \rangle - \langle \bar{Y}_i, \overline{[X, Y_j]} \rangle \\
&= X(\delta_{ij}) \\
&= 0 ,
\end{aligned}
$$

so $L_X g = 0$.

Let L be a leaf of $\mathcal{F}$. For any $x \in L$ there is an open neighbourhood V_x of 0 in $\mathbb{R}^q$ on which the map given by

$$T_x(t) = (e^{t_1 Y_1} \circ \cdots \circ e^{t_q Y_q})(x) , \qquad t = (t_1, \ldots, t_q) ,$$

is defined. In fact we have

$$\left. \frac{\partial T_x(t)}{\partial t_i} \right|_{t=0} = (Y_i)_x ,$$

hence $(dT_x)_0$ is injective. Therefore we may choose V_x so small that T_x is an embedding and that $T_x(V_x)$ is a transversal section of $(M, \mathcal{F})$ at x. Now let $\gamma \colon [0, 1] \to L$ be a curve in L, and put $\Gamma = \gamma([0, 1])$. By compactness of Γ we can find an open neighbourhood V of 0 in $\mathbb{R}^q$ such that for any $x \in \Gamma$, the map T_x embeds V as a transversal section $S_x = T_x(V)$ in M. Since the vector fields $Y_1, \ldots, Y_q$ are projectable, their flows preserve the foliation, therefore for any $t \in V$ the set $\{T_x(t) \mid x \in \Gamma\}$ is contained in a leaf of $\mathcal{F}$. Thus $\mathrm{hol}^{S_{\gamma(1)} S_{\gamma(0)}}(\gamma)(T_{\gamma(0)}(t)) = T_{\gamma(1)}(t)$, and in particular $\mathrm{hol}(\gamma) = \mathrm{id}$ when γ is a loop. $\qquad\square$

Theorem 4.8 *Let $\mathcal{F}$ be a transversely parallelizable foliation of a connected manifold M, for which we can choose complete projectable vector fields $Y_1, \ldots, Y_q$ such that $\bar{Y}_1, \ldots, \bar{Y}_q$ form a transverse parallelism for $(M, \mathcal{F})$. Then $(M, \mathcal{F})$ is homogeneous. In particular, any transversely parallelizable foliation on a compact connected manifold is homogeneous.*

Proof Since $Y_1, \ldots, Y_q$ are complete, the function given in the proof of Proposition 4.7

$$T_x(t) = T(x, t) = (e^{t_1 Y_1} \circ \cdots \circ e^{t_q Y_q})(x) , \qquad t = (t_1, \ldots, t_q) ,$$

is well-defined on $M \times V$ for an open neighbourhood V of 0 in $\mathbb{R}^q$. The map $T(\,\text{-}\,, t)$ is an automorphism of the foliated manifold $(M, \mathcal{F})$, for any $t \in V$. Let L be a leaf of $\mathcal{F}$ and $x \in L$, and denote by A the set of those points y of M for which there exists an automorphism of $(M, \mathcal{F})$ which maps x to y. The set A is saturated by Exercise 4.2 (2). The existence of the automorphisms $T(\,\text{-}\,, t)$ and the fact that T_x is (locally around 0) an embedding transversal to the leaves imply that A is an open (saturated) neighbourhood of L. The connectedness of M now yields that $(M, \mathcal{F})$ is homogeneous.   □

Since we know now that a transversely parallelizable foliation $\mathcal{F}$ of a compact manifold M is homogeneous, we may consider the basic foliation $\mathcal{F}_{\text{bas}}$ of M associated to $\mathcal{F}$ (Subsection 4.1.1).

Theorem 4.9 *Let $\mathcal{F}$ be a transversely parallelizable foliation of a compact manifold M, and let $\mathcal{F}_{\text{bas}}$ be the basic foliation of M associated to $\mathcal{F}$. Then for any leaf L of $\mathcal{F}_{\text{bas}}$ we have*
(i) any basic function on $(L, \mathcal{F}|_L)$ is constant,
(ii) $(L, \mathcal{F}|_L)$ is transversely parallelizable, and
(iii) $l(L, \mathcal{F}|_L)$ is a finite dimensional Lie algebra over $\mathbb{R}$, with

$$\dim l(L, \mathcal{F}|_L) = \operatorname{codim} \mathcal{F} - \operatorname{codim} \mathcal{F}_{\text{bas}} .$$

Proof (i) Take any $x \in L$ and let $\bar{Y}_1, \ldots, \bar{Y}_q$ be a transverse parallelism for $(M, \mathcal{F})$. Put $q' = \operatorname{codim} \mathcal{F}_{\text{bas}}$. We can choose this parallelism so that $(Y_1)_x, \ldots, (Y_{q'})_x$ span a subspace of $T_x(M)$ complementary to $T_x(\mathcal{F}_{\text{bas}})$. Define $T: L \times \mathbb{R}^{q'} \to M$ by

$$T(x, t) = (e^{t_1 Y_1} \circ \cdots \circ e^{t_{q'} Y_{q'}})(x) , \qquad t = (t_1, \ldots, t_{q'}) ,$$

as before. Since $Y_1, \ldots, Y_{q'} \in L(M, \mathcal{F})$ and $L(M, \mathcal{F}) \subset L(M, \mathcal{F}_{\text{bas}})$ (Lemma 4.5), the diffeomorphism $T(\,\text{-}\,, t)$ preserves both foliations, $\mathcal{F}$

and $\mathcal{F}_{\text{bas}}$. Now recall that $Q = M/\mathcal{F}_{\text{bas}}$ is a manifold and that the projection $\pi\colon M \to Q$ is a fibre bundle (Theorem 4.3 (v)). Thus T induces a map $\bar{T}\colon \mathbb{R}^{q'} \to Q$ such that the diagram

$$
\begin{array}{ccc}
L \times \mathbb{R}^{q'} & \xrightarrow{\ T\ } & M \\
{\scriptstyle\text{pr}_2}\Big\downarrow & & \Big\downarrow{\scriptstyle\pi} \\
\mathbb{R}^{q'} & \xrightarrow{\ \bar{T}\ } & Q
\end{array}
$$

commutes. By our choice of $Y_1, \ldots, Y_{q'}$ it follows that $T(x, \text{-})$ is an embedding on an open neighbourhood V of 0 in $\mathbb{R}^{q'}$ transversal to the leaves of $\mathcal{F}_{\text{bas}}$, and hence $\bar{T}|_V$ is an embedding. In turn this implies that $T|_{L \times V}$ is an embedding as well. Write $U = T(L \times V)$.

Now take any basic function f on $(L, \mathcal{F}|_L)$. Our plan is to extend f to a basic function h on $(M, \mathcal{F})$. Since any such function is constant on the leaves of the basic foliation, this would imply that f is constant. To this end, let g be a function on Q with compact support in $\pi(U)$ such that $g(\pi(x)) = 1$. For any $y \in U$ write $T^{-1}(y) = (a(y), b(y)) \in L \times V$. Now define a function h on U, obviously basic, by

$$
h(y) = f(a(y))g(\pi(y)) \,,
$$

and extend it by 0 to all of M.

(ii) Let $x \in L$. We can choose the transverse parallelism $\bar{Y}_1, \ldots, \bar{Y}_q$ in such a way that $(Y_{q'+1})_x, \ldots, (Y_q)_x$ span a subspace of $T_x(\mathcal{F}_{\text{bas}})$ complementary to $T_x(\mathcal{F})$. By Lemma 4.5 each vector field Y_i is also projectable with respect to $\mathcal{F}_{\text{bas}}$. Since $(Y_i)_x$ is tangent to L for $i = q' + 1, \ldots, q$, the projectability of Y_i with respect to $\mathcal{F}_{\text{bas}}$ implies that the same is true at any point along the leaf L. Therefore $Y_{q'+1}|_L, \ldots, Y_q|_L$ are in $L(L, \mathcal{F}_{\text{bas}}|_L)$ and their projections to $l(L, \mathcal{F}_{\text{bas}}|_L)$ form a transverse parallelism.

(iii) This follows directly from (i) and (ii). □

Example 4.10 The basic foliation of Theorem 4.9 is homogeneous by Theorem 4.3 (ii), but need not be transversely parallelizable, as the following example shows.

The Kronecker foliation $\mathcal{F}$ of the torus T^2 is transversely parallelizable. In fact, if we write (φ, θ) for the coordinates of a point of $T^2 = S^1 \times S^1$, the field $\frac{\partial}{\partial \theta}$ (and also $\frac{\partial}{\partial \varphi}$) is a transverse parallelism for the foliation. The basic foliation associated to $(T^2, \mathcal{F})$ is the trivial foliation of codimension 0.

Take now $M = T^2 \times S^2$, and let $\mathcal{G}$ be the foliation of M of dimension 1 given by the product of the Kronecker foliation of T^2 with the trivial foliation of S^2 of dimension 0. We will show that $(M, \mathcal{G})$ is also transversely parallelizable. For this purpose, identify S^2 with the space of vectors in $\mathbb{R}^3$ of norm 1, and identify the tangent space of S^2 at a point $r \in S^2$ with a subspace of $\mathbb{R}^3$ in the natural way. At a point $r \in S^2$, we can uniquely decompose each of the standard basic vectors e_1, e_2, e_3 of $\mathbb{R}^3$ into a vector parallel to r and one tangent to S^2, thus giving

$$e_1 = a_1(r)r + (v_1)_r \ ,$$

$$e_2 = a_2(r)r + (v_2)_r \ ,$$

$$e_3 = a_3(r)r + (v_3)_r$$

for some functions a_1, a_2, a_3 on S^2 and vector fields v_1, v_2, v_3 on S^2. Now define the vector fields Y_1, Y_2, Y_3 on M by

$$(Y_1)_{(p,r)} = \left(a_1(r) \left(\frac{\partial}{\partial \theta} \right)_p , (v_1)_r \right) ,$$

$$(Y_2)_{(p,r)} = \left(a_2(r) \left(\frac{\partial}{\partial \theta} \right)_p , (v_2)_r \right) ,$$

$$(Y_3)_{(p,r)} = \left(a_3(r) \left(\frac{\partial}{\partial \theta} \right)_p , (v_3)_r \right)$$

for any $(p, r) \in M = T^2 \times S^2$. It is easy to check that Y_1, Y_2, Y_3 form a transverse parallelism for $(M, \mathcal{G})$. The associated basic foliation is given by the fibres of the second projection $\mathrm{pr}_2 \colon T^2 \times S^2 \to S^2$, and this foliation is not transversely parallelizable, since S^2 is not parallelizable (Example 4.4 (2)).

4.2 Principal bundles

This section consists of two parts. In the first part, 4.2.1, we will briefly summarize some standard material concerning connections on principal G-bundles on smooth manifolds. There are many extensive treatments of this subject in the literature, for example Kobayashi–Nomizu (1963) or Dupont (1978). In the second part, we will give a parallel treatment of principal G-bundles and connections on the 'space' of leaves of a foliation.

4.2.1 Connections on principal bundles

Let G be a Lie group and M a manifold. A *principal G-bundle* on M (the base space) is a manifold E (the total space) together with a surjective submersion $\pi\colon E \to M$ and a right G-action $\mu\colon E \times G \to E$ on the fibres (i.e. $\pi \circ \mu = \pi \circ \mathrm{pr}_1$) for which the map $(\mu, \mathrm{pr}_1)\colon E \times G \to E \times_M E$ is a diffeomorphism. As usual we shall write $\mu(e, g) = eg = R_g(e)$, and denote the principal G-bundle by (E, π, μ) or simply by E.

An example is the trivial G-bundle $\mathrm{pr}_1\colon M \times G \to M$ on M, where the action is given by $(e, g)g' = (e, gg')$. Note that the G-action on a principal G-bundle E is free and transitive along the fibres of π (hence $M = E/G$) and that $\pi\colon E \to M$ is locally isomorphic to the trivial G-bundle, i.e. every point of M has an open neighbourhood U and an equivariant diffeomorphism $\phi\colon \pi^{-1}(U) \to U \times G$ over U. In fact, one may use this as an alternative definition of a principal G-bundle.

Let E be a principal G-bundle on M and E' a principal G-bundle on M'. A *bundle map* from E' to E is an equivariant map $f\colon E' \to E$. Such a bundle map induces a map $\bar{f}\colon M' \to M$ with $\pi \circ f = \bar{f} \circ \pi'$. We say that f is a bundle map over $\bar{f}$. A bundle map $f\colon E' \to E$ is an isomorphism of principal G-bundles if and only if the induced map $\bar{f}$ is a diffeomorphism (exercise). Let $h\colon M' \to M$ be any smooth map. Then the pull-back $h^*E = M' \times_M E$ is also a principal G-bundle on M' with respect to the natural action. Note that a bundle map $f\colon E' \to E$ induces and isomorphism of principal G-bundles $E' \to \bar{f}^*E$ over the identity, by the exercise above.

Exercise 4.11 A principal G-bundle on M is called *trivial* if it is isomorphic to the (trivial) principal G-bundle $M \times G$ on M. Show that a principal G-bundle E on M is trivial if and only if it admits a global section, i.e. a map $\sigma\colon M \to E$ satisfying $\pi \circ \sigma = \mathrm{id}$.

Example 4.12 Let M be a manifold of dimension n and let $F(M)$ be the frame bundle of M (see also Section 2.4). Recall that $F(M)$ is a smooth fibre bundle on M for which the fibre $F_x(M) = \pi^{-1}(x)$ is the manifold of all ordered bases of the tangent space $T_x(M)$, i.e. of all isomorphisms $e\colon \mathbb{R}^n \to T_x(M)$. The frame bundle admits a canonical right action of the Lie group $GL(n, \mathbb{R})$ given by composition, i.e. $eA = e \circ A$ for $e \in F_x(M)$ and $A \in GL(n, \mathbb{R})$. This action makes the frame bundle $F(M)$ into a principal $GL(n, \mathbb{R})$-bundle on M. Any local diffeomorphism $f\colon M \to M'$ induces a bundle map $F(f)\colon F(M) \to F(M')$ by $F(f)(e) = (df)_{\pi(e)} \circ e$.

Note that there is a natural 1-form $\theta \in \Omega^1(F(M), \mathbb{R}^n)$, called the *canonical form*, given by

$$\theta_e = e^{-1} \circ (d\pi)_e \ .$$

This form has the following properties:

(i) $\text{Ker}(\theta_e) = \text{Ker}(d\pi)_e$, and

(ii) θ is equivariant, in the sense that $R_A^* \theta = A^{-1} \circ \theta$ for any matrix $A \in GL(n, \mathbb{R})$.

Indeed, for any $\xi \in T_e(F(M))$ we have

$$
\begin{aligned}
(R_A^* \theta)_e(\xi) &= \theta_{eA}((dR_A)_e(\xi)) \\
&= (eA)^{-1}((d\pi)_{eA}((dR_A)_e(\xi))) \\
&= A^{-1}(e^{-1}(d(\pi \circ R_A)_e(\xi))) \\
&= A^{-1}(e^{-1}(d(\pi)_e(\xi))) \\
&= A^{-1}(\theta_e(\xi)) \ .
\end{aligned}
$$

Moreover, the canonical form is invariant under diffeomorphisms. In other words, if $f \colon M \to M'$ is a diffeomorphism, then

$$F(f)^* \theta' = \theta \ ,$$

where θ' is the canonical form on $F(M')$ (exercise).

Now if M is a Riemannian manifold, one considers the orthogonal frame bundle $OF(M)$, which is the subbundle of $F(M)$ made up out of the orthogonal isomorphisms $e \colon \mathbb{R}^n \to T_x(M)$. The orthogonal group $O(n)$ leaves $OF(M)$ invariant, furthermore $OF(M)$ is a principal $O(n)$-bundle on M. If $f \colon M' \to M$ is an isometry, the map $F(f)$ restricts to a bundle map of principal $O(n)$-bundles $OF(f) \colon OF(M') \to OF(M)$. The restriction of the canonical form to $OF(M)$, called the *canonical form on* $OF(M)$, is equivariant with respect to the actions of $O(n)$ on $OF(M)$ and on $\mathbb{R}^n$. The canonical form on $OF(M)$ is preserved by isometries.

Let E be a principal G-bundle on M, and let $\mathfrak{g}$ be the Lie algebra of the Lie group G. A *connection* on E is a 1-form ω on E with values in $\mathfrak{g}$ satisfying

(i) $\omega_e \circ \nu_e = \text{id}$, where $\nu_e \colon \mathfrak{g} \to T_e(E)$ is the differential at 1 of the map $G \to E$ which sends g to eg, and

(ii) $R_g^* \omega = \text{Ad}(g^{-1}) \circ \omega$, where $R_g \colon E \to E$ is given by $R_g e = eg$.

A connection ω on E gives us a subspace $H_e = \text{Ker}(\omega_e)$ of $T_e(E)$ at any point $e \in E$, which is called the subspace of *horizontal vectors*. It is complementary (by (i)) to the subspace of *vertical vectors*, given by $V_e = \text{Ker}(d\pi)_e = \nu_e(\mathfrak{g})$ (and hence independent of the connection). Condition (ii) is equivalent to $(dR_g)_e(H_e) = H_{eg}$.

The *curvature form* associated to the connection ω on E is a 2-form $\Omega \in \Omega^2(E, \mathfrak{g})$ on E with values in the Lie algebra $\mathfrak{g}$ of the Lie group G, given by

$$\Omega = d\omega + \frac{1}{2}[\omega, \omega] \ .$$

The connection ω is called *flat* if $\Omega = 0$.

If $f \colon E' \to E$ is a bundle map of principal G-bundles and ω is a connection on E with curvature Ω, then $f^*\omega$ is a connection on E' with curvature $f^*\Omega$ (exercise).

Example 4.13 The trivial principal G-bundle $G \to 1$ over the one point space has a unique connection $\omega_g = (dL_{g^{-1}})_g$, where $L_g \colon G \to G$ is the left translation $L_g(g') = gg'$. More generally, the trivial principal G-bundle $M \times G$ on M has a canonical connection ω obtained as the pull-back of the unique connection on $G \to 1$ along the projection map $\text{pr}_2 \colon M \times G \to G$,

$$\omega_{(x,g)} = (dL_{g^{-1}})_g \circ (d\pi_2)_{(x,g)} \ .$$

This connection is flat.

Exercise 4.14 Let $\alpha \colon G \to H$ be a homomorphism of Lie groups and let E be a principal G-bundle on M.

(i) Define a right G-action on $E \times H$ by $(e, h)g = (eg, \alpha(g^{-1})h)$, and let $E \times_G H$ be the space of orbits $(E \times H)/G$ of this action. Show that $E \times_G H$ is a principal H-bundle on M with respect to the projection $\pi'((e, h)G) = \pi(e)$ and the action $((e, h)G)h' = (e, hh')G$.

(ii) Note that the map $f \colon E \to E \times_G H$, given by $f(e) = (e, 1)G$, is equivariant in the sense that $f(eg) = f(e)\alpha(h)$.

(iii) Show that for any connection ω on E, there exists a unique connection ω' on $E \times_G H$ satisfying $f^*\omega' = (d\alpha)_1 \circ \omega$.

(iv) If Ω and Ω' are the curvature forms of ω and ω' respectively, then we have $f^*\Omega' = (d\alpha)_1 \circ \Omega$.

(v) Let E' be a principal H-bundle on M with an equivariant map $f' \colon E \to E'$ over M. In this case the principal H-bundle E' is called an *extension* of E (or we say that E' is *induced* by E) with respect to

α. Show that for any such extension E' there exists an isomorphism of principal H-bundles $\phi \colon E \times_G H \to E'$ satisfying $f' = \phi \circ f$.

REMARK. From this exercise, one sees that if a principal H-bundle E' is an extension of a principal bundle for a discrete group, the bundle E' has a flat connection. In fact, the converse is also true, by the following proposition:

Proposition 4.15 *Let E be a principal G-bundle on M and let ω be a connection on E. Then $\mathrm{Ker}(\omega)$ is an integrable subbundle of $T(E)$ if and only if ω is flat. If M is connected and ω is flat, then the restriction of the projection $E \to M$ to any leaf $\tilde{M}$ of the foliation on E given by $\mathrm{Ker}(\omega)$ is a principal $G_{\tilde{M}}^{\delta}$-bundle on M, where $G_{\tilde{M}}^{\delta}$ is the isotropy group $G_{\tilde{M}} \subset G$ of $\tilde{M}$ equipped with the discrete topology. Furthermore, the bundle E is the extension of $\tilde{M}$ with respect to $G_{\tilde{M}}^{\delta} \to G$.*

REMARK. Note that if $\mathrm{Ker}(\omega)$ is integrable, then the foliation given by $\mathrm{Ker}(\omega)$ on E is invariant under the action of G. The group $G_{\tilde{M}}$ is, up to conjugation in G, independent of the choice of the leaf $\tilde{M}$, and it is referred to as the *holonomy group* of the (flat) connection ω. The reader should be careful in distinguishing between this notion and the one in Section 2.1. Note also that in this case the extension $E = \tilde{M} \times_{G_{\tilde{M}}^{\delta}} G$ is a flat bundle (Section 1.3).

Proof (of Proposition 4.15) First assume that ω is flat. The kernel of ω is a subbundle of $T(E)$ because ω_e is surjective for any $e \in E$. If X and Y are sections of $\mathrm{Ker}(\omega)$, we have

$$
\begin{aligned}
\omega([X,Y]) &= X(\omega(Y)) - Y(\omega(X)) - 2d\omega(X,Y) \\
&= X(\omega(Y)) - Y(\omega(X)) + [\omega(X), \omega(Y)] \\
&= 0\,,
\end{aligned}
$$

so $[X,Y]$ is also a section of $\mathrm{Ker}(\omega)$. Therefore $\mathrm{Ker}(\omega)$ is involutive, and defines a foliation of E. Assume that M is connected, and let $\tilde{M}$ be a leaf of the foliation given by $\mathrm{Ker}(\omega)$. The restriction of π to $\tilde{M}$ is clearly a local diffeomorphism because $(d\pi)_e$ restricted to the subspace of horizontal vectors H_e is an isomorphism. Let $G_{\tilde{M}}$ be the isotropy group of $\tilde{M}$, i.e. $G_{\tilde{M}} = \{g \in G \mid \tilde{M}g = \tilde{M}\}$. Let $G_{\tilde{M}}^{\delta}$ be the group $G_{\tilde{M}}$ equipped with the discrete topology. Then $\tilde{M} \to M$ is a principal $G_{\tilde{M}}^{\delta}$-bundle over M, hence a covering projection. The fact that E is isomorphic to $\tilde{M} \times_{G_{\tilde{M}}^{\delta}} G$ follows from Exercise 4.14 (v).

To prove the converse, assume that $\text{Ker}(\omega)$ is an involutive subbundle of $T(E)$. From

$$2\Omega(X, Y) = X(\omega(Y)) - Y(\omega(X)) - \omega([X, Y]) + [\omega(X), \omega(Y)]$$

it is clear that $\Omega(X, Y) = 0$ if X and Y are horizontal, i.e. if they are sections of $\text{Ker}(\omega)$. But since $\Omega_e(\xi, \zeta) = 0$ if either of the tangent vectors $\xi, \zeta \in T_e(E)$ is vertical (this is not very difficult to see in this special case where $\text{Ker}(\omega)$ is involutive; but in fact, it is a general property of curvature, see Kobayashi–Nomizu (1963) or Dupont (1978)), we see that ω is flat. $\square$

Proposition 4.16 *Any principal G-bundle admits a connection.*

Proof Let E be a principal G-bundle on M, and let (U_i) be a cover of M such that $E|_{U_i} = \pi^{-1}(U_i)$ is isomorphic to the trivial principal G-bundle. In particular, each $E|_{U_i}$ has a connection ω_i by Example 4.13. Now let (α_i) be a partition of unity subordinate to (U_i). Define a connection on E by $\omega = \sum_i \alpha_i \omega_i$. $\square$

Now let M be a manifold and consider the principal $GL(n, \mathbb{R})$-bundle of frames $F(M)$ on M. Recall that we have the canonical form θ on $F(M)$ with values in $\mathbb{R}^n$ satisfying $\text{Ker}(\theta_e) = \text{Ker}(d\pi)_e$. For a connection ω on $F(M)$ we define the *torsion form* $\Theta \in \Omega^2(F(M), \mathbb{R}^n)$ by

$$\Theta = d\theta + \omega \wedge \theta \, .$$

Here $\omega \wedge \theta$ is given by $(\omega \wedge \theta)_e(\xi, \zeta) = \frac{1}{2}(\omega_e(\xi)\theta_e(\zeta) - \omega_e(\zeta)\theta_e(\xi))$. The connection ω is *torsion free* if $\Theta = 0$. The same observations and definitions apply to the orthogonal frame bundle $OF(M)$ if M is a Riemannian manifold.

Theorem 4.17 *For a Riemannian manifold M, there is a unique torsion free connection on $OF(M)$.*

REMARK. For the proof, see Kobayashi–Nomizu (1963). The unique torsion free connection on $OF(M)$ is called the *Levi-Città connection*. It can be uniquely extended to a connection on $F(M)$, which is called the Levi-Città connection (see Exercise 4.14) as well. Note that if $f: M' \to M$ is an isometry between Riemannian manifolds and if ω is the Levi-Città connection on $F(M)$, then $F(f)^*\omega$ is the Levi-Città connection on $F(M')$, and $OF(f)^*\omega$ is the Levi-Città connection on $OF(M')$.

Corollary 4.18 *For any manifold M, the manifold $F(M)$ is paralleliz-able. For a Riemannian manifold M, the manifolds $F(M)$ and $OF(M)$ admit canonical parallelisms invariant under isometries.*

Proof Let ω be any connection on $F(M)$, and put

$$\tau_e = (\theta_e, \omega_e) \colon T_e(F(M)) \longrightarrow \mathbb{R}^n \times \mathfrak{gl}(n, \mathbb{R}) \ .$$

Note that τ_e is an isomorphism because $\mathrm{Ker}(\theta_e) \oplus \mathrm{Ker}(\omega_e) = V_e \oplus H_e = T_e(F(M))$. Let $e_1, \dots, e_n$ be the standard basis of $\mathbb{R}^n$ and let

$$E_{11}, E_{12}, \dots, E_{nn}$$

be the standard basis of $\mathfrak{gl}(n, \mathbb{R})$. Then define vector fields

$$Y_1, \dots, Y_n, Z_{11}, Z_{12}, \dots, Z_{nn}$$

on $F(M)$ by $(Y_i)_e = \tau_e^{-1}(e_i, 0)$ and $(Z_{ij})_e = \tau_e^{-1}(0, E_{ij})$. This is a parallelism on $F(M)$. If M is Riemannian, we can choose ω to be the Levi-Città connection on $F(M)$. To prove that $OF(M)$ is parallelizable one uses the same argument, with $\mathfrak{gl}(n, \mathbb{R})$ replaced by $\mathfrak{o}(n)$. The paral-lelisms obtained by using the Levi-Città connection are invariant under isometries because the canonical form and the Levi-Città connection are both invariant under isometries.   □

4.2.2 Transverse principal bundles

Let $(M, \mathcal{F})$ be a foliated manifold and let G be a Lie group. A *transverse principal G-bundle* on $(M, \mathcal{F})$ is a principal G-bundle E on M, equipped with a foliation $\tilde{\mathcal{F}}$ such that

(i) $\tilde{\mathcal{F}}$ is preserved by the action of G, and
(ii) the projection $\pi \colon E \to M$ maps each leaf $\tilde{L}$ of $\tilde{\mathcal{F}}$ onto a leaf $L = \pi(\tilde{L})$ of $\mathcal{F}$, and the restriction of π to $\tilde{L}$ is a covering projection $\tilde{L} \to L$ which is a quotient of the holonomy cover of the leaf L of $\mathcal{F}$.

REMARK. Note that condition (ii) implies that $\dim \tilde{\mathcal{F}} = \dim \mathcal{F}$. A *foliated* principal G-bundle on $(M, \mathcal{F})$ (see Kamber–Tondeur (1975) and Molino (1988)) is a principal G-bundle E on M, equipped with a foliation $\tilde{\mathcal{F}}$ of the same dimension as $\mathcal{F}$ such that $\tilde{\mathcal{F}}$ is preserved by the action of G, while $\pi \colon (E, \tilde{\mathcal{F}}) \to (M, \mathcal{F})$ is a map of foliated manifolds and $T_e(\tilde{\mathcal{F}}) \cap \mathrm{Ker}(d\pi)_e = 0$ for any $e \in E$. It is easy to see that for such a foliated principal G-bundle $(E, \tilde{\mathcal{F}})$, the projection $\pi \colon E \to M$ maps any leaf $\tilde{L}$ of $\tilde{\mathcal{F}}$ onto a leaf $L = \pi(\tilde{L})$ of $\mathcal{F}$, and that the restriction of π to $\tilde{L}$ is a covering projection $\tilde{L} \to L$. However, this covering projection need not be a quotient of the holonomy cover of L. Any transverse principal G-bundle is therefore a foliated principal G-bundle, but not conversely. (See also Example 5.36.)

Let $(E, \tilde{\mathcal{F}})$ be a transverse principal G-bundle on $(M, \mathcal{F})$, and let $\mathfrak{g}$ be the Lie algebra of G. A *projectable connection* on $(E, \tilde{\mathcal{F}})$ is a connection $\omega \in \Omega^1(E, \mathfrak{g})$ on E which satisfies

(i) $T_e(\tilde{\mathcal{F}}) \subset \mathrm{Ker}(\omega_e)$ for any $e \in E$, and
(ii) $L_X \omega = 0$ for any $X \in \mathfrak{X}(\tilde{\mathcal{F}})$.

Here $L_X \omega(Y) = X(\omega(Y)) - \omega([X, Y])$ is the Lie derivative of ω.

We shall now describe the most important example of a transverse principal G-bundle, the transverse frame bundle of a foliation.

Example 4.19 (The transverse frame bundle) Let $\mathcal{F}$ be a foliation of a manifold M. The frame bundle of the normal bundle $N(\mathcal{F})$ of $\mathcal{F}$ is referred to as the *transverse frame bundle* on $(M, \mathcal{F})$, and is denoted by $F(M, \mathcal{F})$. Recall that a point of $F_x(M, \mathcal{F})$ is a linear isomorphism $e \colon \mathbb{R}^q \to N_x(\mathcal{F})$, where q is the codimension of $\mathcal{F}$. The transverse frame bundle is a principal $GL(q, \mathbb{R})$-bundle on M, where the action is given by the composition $eA = e \circ A$, for any $A \in GL(q, \mathbb{R})$.

Let $(s_i \colon U_i \to \mathbb{R}^q)$ be a Haefliger cocycle for $\mathcal{F}$. Each s_i induces a bundle map over s_i between principal $GL(q, \mathbb{R})$-bundles

$$\tilde{s}_i \colon F(M, \mathcal{F})|_{U_i} \longrightarrow F(s_i(U_i)) \subset F(\mathbb{R}^q) = \mathbb{R}^q \times GL(q, \mathbb{R}) \ ,$$

given for any $x \in U_i$ and any $e \in F_x(M, \mathcal{F})$ by

$$\tilde{s}_i(e) = (ds_i)_x \circ e \ .$$

Moreover, if we write $s_{ij} \colon s_j(U_i \cap U_j) \to s_i(U_i \cap U_j)$ for the diffeomorphism satisfying $s_{ij} \circ s_j = s_i$, the diffeomorphism of principal G-bundles $F(s_{ij}) \colon F(s_j(U_i \cap U_j)) \to F(s_i(U_i \cap U_i))$ satisfies $F(s_{ij}) \circ \tilde{s}_j = \tilde{s}_i$. This means that if we write $\tilde{U}_i = F(M, \mathcal{F})|_{U_i}$, then

$$(\tilde{s}_i \colon \tilde{U}_i \longrightarrow \mathbb{R}^q \times GL(q, \mathbb{R}))$$

is a Haefliger cocycle on $F(M, \mathcal{F})$. The foliation on $F(M, \mathcal{F})$ given by this cocycle is called the *lifted foliation* and will be denoted by $\tilde{\mathcal{F}}$. This foliation is preserved by the action of $GL(q, \mathbb{R})$. Moreover the projection $\pi \colon (F(M, \mathcal{F}), \tilde{\mathcal{F}}) \to (M, \mathcal{F})$ is a map of foliated manifolds and the restriction of π to a leaf $\tilde{L}$ of $\tilde{\mathcal{F}}$ is a covering projection onto the corresponding leaf $L = \pi(\tilde{L})$ of $\mathcal{F}$. Furthermore, the group of covering transformations of the covering projection $\tilde{L} \to L$ is exactly the linear holonomy group of L. In particular, $(F(M,\mathcal{F}),\tilde{\mathcal{F}})$ is a transverse principal $GL(q,\mathbb{R})$-bundle on $(M,\mathcal{F})$.

There is a natural 1-form $\theta \in \Omega^1(F(M,\mathcal{F}),\mathbb{R}^q)$, called the *transverse canonical form*, given by

$$\theta_e(\xi) = e^{-1}(\overline{(d\pi)_e(\xi)})$$

for any point $e\colon \mathbb{R}^q \to N_x(\mathcal{F})$ of $F(M,\mathcal{F})$ and any $\xi \in T_e(F(M,\mathcal{F}))$, where $\overline{(d\pi)_e(\xi)}$ denotes the natural projection of $(d\pi)_e(\xi)$ onto $N_x(\mathcal{F})$. With respect to the Haefliger cocycle (s_i) for $\mathcal{F}$ as above, it is easy to check that

$$\theta|_{\tilde{U}_i} = \tilde{s}_i^* \theta_i\ ,$$

where θ_i is the canonical form on the frame bundle $F(s_i(U_i))$. Furthermore, the transverse canonical form in equivariant, i.e. $R_A^* \theta = A^{-1} \circ \theta$ for any $A \in GL(q,\mathbb{R})$, and satisfies $\mathrm{Ker}(\theta_e) = \mathrm{Ker}(d\pi)_e \oplus T_e(\tilde{\mathcal{F}})$.

Now assume that $(\mathcal{F},g)$ is Riemannian foliation of the manifold M. Then we may consider the *transverse orthogonal frame bundle* which is the subbundle of $F(M,\mathcal{F})$ made up out of the orthogonal isomorphisms $e\colon \mathbb{R}^q \to N_x(\mathcal{F})$. It is a principal $O(q)$-bundle on M. We may choose the Haefliger cocycle $(s_i\colon U_i \to \mathbb{R}^q)$ so that $g|_{U_i}$ is the pull-back of a Riemannian metric on $s_i(U_i)$ along s_i (Remark 2.7 (2)). In particular, any diffeomorphism s_{ij} is an isometry. Then the induced map $F(s_{ij})$ restricts to an isomorphism of principal $O(q)$-bundles $OF(s_j(U_i \cap U_j)) \to OF(s_i(U_i \cap U_i))$. In particular, the restrictions of $(\tilde{s}_i)$ to $(OF(M,\mathcal{F})|_{U_i})$ form a Haefliger cocycle on $OF(M,\mathcal{F})$. The associated foliation of $OF(M,\mathcal{F})$ is the restriction of $\tilde{\mathcal{F}}$ and will be denoted again by $\tilde{\mathcal{F}}$. With this, $(OF(M,\mathcal{F}),\tilde{\mathcal{F}})$ is a transverse principal $O(q)$-bundle on $(M,\mathcal{F})$. The restriction of the canonical form to $OF(M,\mathcal{F})$, called the *transverse canonical form* on $OF(M,\mathcal{F})$, is equivariant with respect to the actions of $O(n)$ on $OF(M,\mathcal{F})$ and on $\mathbb{R}^n$.

Let ω_i be the Levi-Cività connection on the frame bundle $F(s_i(U_i))$. Since the Levi-Cività connection is preserved by isometries, we have $\omega_j|_{F(s_j(U_i \cap U_j))} = F(s_{ji})^* \omega_i$. As a consequence, we can define a connection ω on $F(M,\mathcal{F})$ as the amalgamation of the forms

$$\omega|_{\tilde{U}_i} = \tilde{s}_i^* \omega_i\ .$$

It is easy to check that ω is a projectable connection on the transverse principal $GL(q, \mathbb{R})$-bundle of transverse frames $(F(M, \mathcal{F}), \tilde{\mathcal{F}})$, called the *transverse* Levi-Cività connection. This connection restricts to a projectable connection on $(OF(M, \mathcal{F}), \tilde{\mathcal{F}})$. Note that we have

$$\mathrm{Ker}(\theta_e)/T_e(\tilde{\mathcal{F}}) \oplus \mathrm{Ker}(\omega_e)/T_e(\tilde{\mathcal{F}}) = T_e(F(M, \mathcal{F}))/T_e(\tilde{\mathcal{F}}) = N_e(\tilde{\mathcal{F}}) \ .$$

Theorem 4.20 *Let $(\mathcal{F}, g)$ be a Riemannian foliation on M. Then the lifted foliation $\tilde{\mathcal{F}}$ of the transverse frame bundle $F(M, \mathcal{F})$, and its restriction to the transverse orthogonal frame bundle $OF(M, \mathcal{F})$, are transversely parallelizable.*

Proof We shall use the same notation as in Example 4.19. Corollary 4.18 implies that the manifolds $F(s_i(U_i))$ admit canonical parallelisms $Y_1^i, \ldots, Y_q^i, Z_{11}^i, \ldots, Z_{qq}^i$ invariant under isometries s_{ij}. Since $\tilde{s}_i$ induces an isomorphism

$$\hat{s}_i \colon N_e(\tilde{\mathcal{F}}) \longrightarrow T_{\tilde{s}_i(e)}(F(s_i(U_i))) \ ,$$

we may define sections $\bar{Y}_1, \ldots, \bar{Y}_q, \bar{Z}_{11}, \ldots, \bar{Z}_{qq}$ of the normal bundle $N(\tilde{\mathcal{F}})$ by

$$(\bar{Y}_k)_e = \hat{s}_i^{-1}((Y_k^i)_{\tilde{s}_i(e)})$$

and

$$(\bar{Z}_{kl})_e = \hat{s}_i^{-1}((Z_{kl}^i)_{\tilde{s}_i(e)})$$

for any $e \in F(M, \mathcal{F})$. Here i is any index such that $\pi(e) \in U_i$. This definition is independent of the chosen i by the invariance under isometries just mentioned. It is obvious that the sections $\bar{Y}_1, \ldots, \bar{Y}_q, \bar{Z}_{11}, \ldots, \bar{Z}_{qq}$ are transverse vector fields on $(F(M, \mathcal{F}), \tilde{\mathcal{F}})$, because they can be projected along the submersions $\tilde{s}_i$ which define the foliation $\tilde{\mathcal{F}}$. It follows that these sections $\bar{Y}_1, \ldots, \bar{Y}_q, \bar{Z}_{11}, \ldots, \bar{Z}_{qq}$ form a transverse parallelism on $(F(M, \mathcal{F}), \tilde{\mathcal{F}})$. To prove that $(OF(M, \mathcal{F}), \tilde{\mathcal{F}})$ is transversely parallelizable one uses the same argument, with $F(s_i(U_i))$ replaced by $OF(s_i(U_i))$. □

4.3 Lie foliations and Molino's theorem

In this section, we consider so-called 'Lie' foliations, defined by suitable forms with values in a Lie algebra. Such foliations are transversely parallelizable. For each such foliation, one can construct a flat (transverse) principal bundle, whose leaves are known as Darboux covers of M. This construction will show that for a transversely parallelizable foliation, the associated basic foliation is obtained simply by taking closures of leaves. In the last part we show how the results of Sections 4.1, 4.2 and 4.3 can be summarized into a statement known as Molino's theorem for Riemannian foliations.

4.3.1 Lie foliations

Let $\mathfrak{g}$ be a Lie algebra and M a manifold. A *Maurer–Cartan form* with values in $\mathfrak{g}$ is a differential 1-form $\omega \in \Omega^1(M, \mathfrak{g})$ with trivial formal curvature, i.e.

$$d\omega + \frac{1}{2}[\omega, \omega] = 0 .$$

Here $[\omega, \omega]$ is the differential 2-form on M with values in $\mathfrak{g}$ given by $[\omega, \omega](X, Y) = [\omega(X), \omega(Y)]$. If ω is non-singular, i.e. if $\omega_x \colon T_x(M) \to \mathfrak{g}$ is surjective at any point $x \in M$, then the dimension of $\mathfrak{g}$ is finite and $\mathrm{Ker}(\omega)$ is a subbundle of $T(M)$ of codimension $\dim \mathfrak{g}$. In the proof of Proposition 4.15 we have shown that vanishing of the formal curvature implies that the subbundle $\mathrm{Ker}(\omega)$ is involutive and hence defines a foliation $\mathcal{F}$ on M, with $\mathrm{codim}\, \mathcal{F} = \dim \mathfrak{g}$ and

$$T(\mathcal{F}) = \mathrm{Ker}(\omega) .$$

A *Lie foliation* is a foliation defined in this way by a non-singular Maurer–Cartan form.

Proposition 4.21 *Any Lie foliation is transversely parallelizable.*

Proof Let $\mathcal{F}$ be a foliation given by a non-singular Maurer–Cartan form $\omega \in \Omega^1(M, \mathfrak{g})$. Since $\mathrm{Ker}(\omega) = T(\mathcal{F})$, the form ω induces a map

$$\bar{\omega} \colon N(\mathcal{F}) \longrightarrow \mathfrak{g} ,$$

which is a linear isomorphism $\bar{\omega}_x \colon N_x(\mathcal{F}) \to \mathfrak{g}$ at any point $x \in M$. Now choose a basis $e_1, \ldots, e_q$ for $\mathfrak{g}$, and define the sections $\sigma_1, \ldots, \sigma_q$ of $N(\mathcal{F})$ by

$$\sigma_i(x) = \bar{\omega}_x^{-1}(e_i) .$$

It is clear that these sections form a frame for the normal bundle. Moreover, they are also transverse vector fields on $(M, \mathcal{F})$. To see this, choose any vector fields $Y_1, \ldots, Y_q$ on M with $\sigma_i = Y_i + \mathfrak{X}(\mathcal{F})$, and take any

$X \in \mathfrak{X}(\mathcal{F})$. Then we have

$$
\begin{aligned}
\omega([X, Y_i]) &= X(\omega(Y_i)) - Y_i(\omega(X)) - 2d\omega(X, Y_i) \\
&= X(\omega(Y_i)) - Y_i(\omega(X)) + [\omega(X), \omega(Y_i)] \\
&= 0 ,
\end{aligned}
$$

because $\omega(X) = 0$ and $\omega(Y_i) = e_i$ is constant and hence $X(\omega(Y_i)) = 0$ as well. Therefore $[X, Y_i] \in \mathfrak{X}(\mathcal{F})$. □

4.3.2 The Darboux cover

Let ω be a non-singular Maurer–Cartan form on a connected manifold M with values in a Lie algebra $\mathfrak{g}$, and let $\mathcal{F}$ be the corresponding Lie foliation of M. Let G be the unique connected and simply connected Lie group such that $\mathfrak{g}$ is the Lie algebra associated to G (see Serre (1965) or Warner (1983)). We can extend ω to a connection η on the trivial principal G-bundle $\mathrm{pr}_1\colon M \times G \to M$, by setting

$$
\eta_{(x,g)}(\xi, \zeta) = \mathrm{Ad}(g^{-1})\omega_x(\xi) + (dL_{g^{-1}})_g(\zeta) ,
$$

for any $(\xi, \zeta) \in T_{(x,g)}(M \times G) = T_x(M) \oplus T_g(G)$. It is easy to check that $\mathrm{Ker}(\eta)$ is an involutive subbundle of $T(M \times G)$, or equivalently, that η is flat (Proposition 4.15). Therefore we have a foliation $\mathcal{G}$ on $M \times G$, which is invariant under the G-action, with $T(\mathcal{G}) = \mathrm{Ker}(\eta)$.

For any $(x, g) \in M \times G$, the tangent space $T_{(x,g)}(\mathcal{G})$ consists of those tangent vectors (ξ, ζ) for which $\eta_{(x,g)}(\xi, \zeta) = 0$, i.e.

$$
\omega_x(\xi) + (dR_{g^{-1}})_g(\zeta) = 0 .
$$

Therefore $(d\mathrm{pr}_2)_{(x,g)}$ restricts to an epimorphism $T_{(x,g)}(\mathcal{G}) \to T_g(G)$, because ω_x is surjective. Thus if $\tilde{M}$ is a leaf of $\mathcal{G}$, then the restriction of pr_2 to $\tilde{M}$ is a submersion. We shall write $f_\omega\colon \tilde{M} \to G$ for this restriction. Denote by H_ω the holonomy group $G_{\tilde{M}}$ of η, in other words $H_\omega = \{g \in G \mid \tilde{M}g = \tilde{M}\}$, and let H_ω^δ be the group H_ω equipped with the discrete topology. The group H_ω is also called the *holonomy group* of the Lie foliation given by ω. Note that there is no danger of confusing this group with the holonomy group of a leaf of $\mathcal{F}$ because any leaf of $\mathcal{F}$ has trivial holonomy (Proposition 4.7 and Proposition 4.21).

The restriction of pr_1 to $\tilde{M}$, which we shall denote by $\pi\colon \tilde{M} \to M$, is a principal H_ω^δ-bundle, hence a covering projection (Proposition 4.15). Let $\bar{H}_\omega$ be the Lie subgroup of G obtained as the closure of H_ω in G. Then there is a manifold structure on the set of left cosets $G/\bar{H}_\omega$ of $\bar{H}_\omega$

such that the projection $G \to G/\bar{H}_\omega$ is a submersion (see Serre (1965) or Warner (1983)). Since f_ω is an equivariant submersion, it induces a submersion $M \to G/\bar{H}_\omega$.

Next note that $(\xi, 0) \in T_{(x,g)}(\tilde{M})$ if and only if $\xi \in \text{Ker}(\omega_x) = T_x(\mathcal{F})$. This implies that $\pi^*\mathcal{F}$ is the foliation given by the submersion f_ω. In particular, the map π maps the leaves of $\pi^*\mathcal{F}$ diffeomorphically to the leaves of $\mathcal{F}$.

To summarize, for a leaf $\tilde{M}$ of $\mathcal{G}$ we have a diagram, known as the *Darboux cover* of ω,

$$
\begin{array}{ccc}
(\tilde{M}, \pi^*\mathcal{F}) & \xrightarrow{\;f_\omega\;} & G \\
\pi \downarrow & & \downarrow \\
(M, \mathcal{F}) & \longrightarrow & G/\bar{H}_\omega
\end{array}
$$

where π is both a principal H_ω^δ-bundle on M and a map of foliated manifolds which restricts to a diffeomorphism on each leaf, and f_ω is an H_ω-equivariant submersion for which the connected components of the fibres are precisely the leaves of $\pi^*\mathcal{F}$.

The fact that $\mathcal{G}$ is invariant under the G-action implies that the Darboux cover is determined, up to an isomorphism, by the Lie foliation and does not depend on the choice of the leaf $\tilde{M}$ of $\mathcal{G}$. We may choose a base-point $x_0 \in M$ and take $\tilde{M}$ to be the leaf with $(x_0, 1) \in \tilde{M}$. Now the action of the fundamental group of M on $\tilde{M}$ by covering transformations gives us a homomorphism of groups

$$h_\omega \colon \pi_1(M, x_0) \longrightarrow G \;,$$

which is also called the *holonomy homomorphism* of ω. By its definition, it satisfies the identity $\tilde{x}\gamma = \tilde{x}h_\omega(\gamma)$ for any $\tilde{x} \in \tilde{M}$ and $\gamma \in \pi_1(M, x_0)$. The image of h_ω is exactly H_ω. The map f_ω is h_ω-equivariant in the sense that

$$f_\omega(\tilde{x}\gamma) = f_\omega(\tilde{x})h_\omega(\gamma) \;,$$

as follows from the previous identity.

REMARK. The foliations $\pi^*\mathcal{F}$, for each leaf $\tilde{M}$ of $\mathcal{G}$, in fact together form a foliation $\tilde{\mathcal{F}}$ on $M \times G$; more precisely,

$$T(\tilde{\mathcal{F}}) = \text{Ker}(d\text{pr}_2) \cap (d\text{pr}_1)^{-1}(T(\mathcal{F})) \;.$$

Then $(M \times G, \tilde{\mathcal{F}})$ is a transverse principal G-bundle on $(M, \mathcal{F})$, and η is a projectable connection on $(M \times G, \tilde{\mathcal{F}})$.

Example 4.22 Let $\omega_1, \ldots, \omega_q$ be closed, (pointwise) linearly independent 1-forms on a manifold M. Then $\omega = (\omega_1, \ldots, \omega_q)$ is a (closed) Maurer–Cartan form with values in the abelian Lie algebra $\mathfrak{g} = \mathbb{R}^q$ (where the Lie bracket is zero). The Lie foliation given by ω is the one described in Example 4.6 (1). Let us consider the Darboux cover of ω. The associated Lie group is again $\mathbb{R}^q$, and we shall identify $T_g(\mathbb{R}^q)$ with $\mathbb{R}^q$ in the canonical way, for any $g \in \mathbb{R}^q$. The connection on $M \times \mathbb{R}^q$ is now given by

$$\eta_{(x,g)}(\xi, v) = \omega_x(\xi) + v \,,$$

for any $\xi \in T_x(M)$ and $g, v \in \mathbb{R}^q$. Choose $x_0 \in M$ and let $\tilde{M}$ be the leaf of the foliation given by $\mathrm{Ker}(\eta)$ with $(x_0, 0) \in \tilde{M}$. Take a (smooth) path $\gamma \colon [0,1] \to M$ with $\gamma(1) = x_0$, and let $\tilde{\gamma}$ be the unique lift of γ in $\tilde{M}$ with $\tilde{\gamma}(1) = (x_0, 0)$. Put $\tilde{x} = \tilde{\gamma}(0)$. With respect to the product $M \times \mathbb{R}^q$ we can write $\tilde{\gamma} = (\gamma, \tau)$ for a path τ in $\mathbb{R}^q$. We have $\tau(1) = 0$ and $\tau(0) = f_\omega(\tilde{x})$. Since the lift $\tilde{\gamma}$ is horizontal, it follows that

$$0 = \eta\left(\frac{d\gamma}{dt}, \frac{d\tau}{dt}\right) = \omega\left(\frac{d\gamma}{dt}\right) + \frac{d\tau}{dt} \,,$$

and therefore

$$f_\omega(\tilde{x}) = \tau(0) = \int_0^1 \omega\left(\frac{d\gamma}{dt}\right) dt = \int_{[0,1]} \gamma^*\omega \,.$$

Thus, if γ is a loop representing an element of $\pi_1(M, x_0)$ we have

$$h_\omega(\gamma) = \int_{[0,1]} \gamma^*\omega \,.$$

The map $h_\omega \colon \pi_1(M, x_0) \to \mathbb{R}^q$ factors through $H_1(M, \mathbb{Z})$ and the last identity above shows that the factorization is exactly the map corresponding to ω as an element of $H^1_{\mathrm{dR}}(M, \mathbb{R}^q)$.

Now assume that M is compact and that $q = 1$, i.e. that we have a foliation on a compact manifold given by a closed form ω. Then the holonomy group of ω is either discrete or dense in $\mathbb{R}$. In the first case the foliation is given by a submersion $M \to S^1$, since the quotient of $\mathbb{R}$ by a discrete subgroup is the circle. If the holonomy group of ω is dense in $\mathbb{R}$, then all the leaves are dense in M. In fact, this is true in general:

Lemma 4.23 *Let ω be a non-singular Maurer–Cartan form on a compact connected manifold M and let $\mathcal{F}$ be the foliation of M given by $\mathrm{Ker}(\omega)$. Then the map $f_\omega \colon \tilde{M} \to G$, defined on the Darboux cover* *of ω, is a fibre bundle with connected fibres. Moreover, the following conditions are equivalent.*

(i) The leaves of $\mathcal{F}$ are dense in M.
(ii) The holonomy group of ω is dense in G.
(iii) Any basic function on $(M, \mathcal{F})$ is constant.

Proof The foliation $\mathcal{F}$ is transversely parallelizable by Proposition 4.21, so we can choose a transverse parallelism $\bar{Y}_1, \ldots, \bar{Y}_q$ for $(M, \mathcal{F})$ such that $\omega(Y_i)$ are constant functions, say $\omega_x((Y_i)_x) = e_i$. The projectable vector fields $Y_1, \ldots, Y_q$ are complete because M is compact. Let $\tilde{Y}_1, \ldots, \tilde{Y}_q$ be the unique lifts of $Y_1, \ldots, Y_q$ in $\tilde{M}$. Explicitly we can describe them by

$$(\tilde{Y}_i)_{(x,g)} = ((Y_i)_x, -(dR_g)_1(e_i)) \in T_{(x,g)}(\tilde{M}) \subset T_x(M) \oplus T_g(G) .$$

The vector fields are complete and projectable with respect to the foliation $\pi^*\mathcal{F}$. In fact, they are projectable along f_ω, their projections to G are right invariant, complete and form a global parallelism of G. Take any $g \in G$ and put $F = f_\omega^{-1}(g)$. We can now use the flows of the vector fields $\tilde{Y}_1, \ldots, \tilde{Y}_q$ to define a map $T: F \times V \to \tilde{M}$ by

$$T_x(t) = T(x, t) = (e^{t_1 \tilde{Y}_1} \circ \cdots \circ e^{t_q \tilde{Y}_q})(x) , \qquad t = (t_1, \ldots, t_q) ,$$

for an open neighbourhood V of 0 in $\mathbb{R}^q$. Since the vector fields $\tilde{Y}_1, \ldots, \tilde{Y}_q$ are projectable along f_ω, the map T induces a map $\tau: V \to G$ with $f_\omega \circ T = \tau \circ \mathrm{pr}_2$. This map τ is in fact defined by the flows of the global parallelism of G mentioned above. We can therefore choose V small enough that τ is an embedding, and hence T gives us a local trivialization of $f_\omega: \tilde{M} \to G$ around g. This proves that f_ω is a fibre bundle over the connected manifold G. In particular, the space of leaves $\tilde{M}/\pi^*\mathcal{F}$ is a connected Hausdorff manifold and the map f_ω factors as a covering projection $\tilde{M}/\pi^*\mathcal{F} \to G$. Since G is simply connected, this covering projection must be a diffeomorphism. Therefore the fibres of f_ω are connected.

Let us now prove that the statements (i), (ii) and (iii) are equivalent. The statement (i) clearly implies the statement (iii). Next, the statement (iii) implies (ii) because if the holonomy group H_ω is not dense in G then f_ω factors as a non-trivial submersion $M \to G/\bar{H}_\omega$, and the pullback of any non-constant function on $G/\bar{H}_\omega$ gives us a non-constant basic function on $(M, \mathcal{F})$. Finally, we shall prove that (ii) implies (i). Assume therefore that H_ω is dense in G, and take any point $x \in M$. Take any open neighbourhood U of x. We have to show that any leaf of $\mathcal{F}$ intersects U. The open subset $\pi^{-1}(U)$ of $\tilde{M}$ is invariant under the action of H_ω, so $f_\omega(\pi^{-1}(U))$ is an H_ω-invariant open subset of G. Since H_ω is dense in G, it follows that $f_\omega(\pi^{-1}(U)) = G$, i.e. any fibre of f_ω intersects $\pi^{-1}(U)$. Since the fibres of f_ω are connected, they are precisely the leaves of $\pi^*\mathcal{F}$. It follows that any leaf of $\mathcal{F}$ intersects U.

$\square$

Theorem 4.24 *Any transversely parallelizable foliation on a compact connected manifold, all of whose basic functions are constant, is a Lie foliation with dense leaves. In fact, it is given by a canonical Maurer–Cartan form with values in the Lie algebra of transverse vector fields.*

Proof Let $(M, \mathcal{F})$ be a transversely parallelizable foliated manifold of codimension q all of whose basic functions are constant. The Lie algebra $l(M, \mathcal{F})$ is a free module of rank q over $\Omega^0_{\text{bas}}(M, \mathcal{F})$ because $\mathcal{F}$ is transversely parallelizable. Hence by the assumption on basic functions, $l(M, \mathcal{F})$ is of dimension q over $\mathbb{R}$. Thus for any point $x \in M$, the evident evaluation map

$$\text{ev}_x \colon l(M, \mathcal{F}) \longrightarrow N_x(\mathcal{F}) \,,$$

sending $\bar{Y}$ to $\bar{Y}_x$, is an isomorphism. We can therefore define a 1-form ω on M with values in $l(M, \mathcal{F})$ by

$$\omega_x(\xi) = \text{ev}_x^{-1}(\bar{\xi}) \,,$$

where $\xi \in T_x(M)$ and $\bar{\xi} \in N_x(\mathcal{F})$ is its projection to the normal bundle. Note that, in particular, $\omega_x(Y_x) = \bar{Y}$ for any vector field $Y \in L(M, \mathcal{F})$. Since clearly we have $\text{Ker}(\omega) = T(\mathcal{F})$, we only need to show that ω is a Maurer–Cartan form, i.e. that the formal curvature of ω vanishes. Observe that any tangent vector on M occurs as the value of a projectable vector field in $L(M, \mathcal{F})$. (This is always true locally, but in our case it is even globally true.) Thus it is sufficient to prove that the curvature vanishes on the projectable vector fields. Take any two projectable vector fields $Y, Z \in L(M, \mathcal{F})$. Since the functions $\omega(Y)$ and $\omega(Z)$ are constant with values $\bar{Y}$ and $\bar{Z}$ respectively, their derivatives vanish and

$$
\begin{aligned}
2\left(d\omega + \frac{1}{2}[\omega, \omega]\right)(Y, Z) \\
= \quad & Y(\omega(Z)) - Z(\omega(Y)) - \omega([Y, Z]) + [\omega(Y), \omega(Z)] \\
= \quad & -\overline{[Y, Z]} + [\bar{Y}, \bar{Z}] \\
= \quad & 0 \,.
\end{aligned}
$$

Finally, the foliation $\mathcal{F}$ has dense leaves by Lemma 4.23. $\square$

Corollary 4.25 *Let $\mathcal{F}$ be a transversely parallelizable foliation on a compact manifold M, and let $\mathcal{F}_{\text{bas}}$ be the associated basic foliation. Then the leaves of $\mathcal{F}_{\text{bas}}$ are the closures of the leaves of $\mathcal{F}$.*

Proof This follows by applying Theorem 4.24 to the restriction of $\mathcal{F}$ to a leaf of $\mathcal{F}_{\text{bas}}$, as described in Theorem 4.9. □

4.3.3 Molino's structure theorem

The results proved in Sections 4.1, 4.2 and 4.3 can now be summarized in the following theorem of Molino.

Theorem 4.26 (Molino's structure theorem) *Let $(\mathcal{F}, g)$ be a Riemannian foliation of a compact connected manifold M, and let $\tilde{\mathcal{F}}$ be the associated lifted foliation of the transverse orthogonal frame bundle $OF(M, \mathcal{F})$.*

(i) The foliated manifold $(OF(M, \mathcal{F}), \tilde{\mathcal{F}})$ is transversely parallelizable.

(ii) There exists a manifold N with an $O(q)$-action and an $O(q)$-equivariant fibre bundle $s: OF(M, \mathcal{F}) \to N$ such that the fibres of s are exactly the closures of the leaves of $\tilde{\mathcal{F}}$.

(iii) The Lie algebra $\mathfrak{g}$ of transverse vector fields of $\tilde{\mathcal{F}}|_{s^{-1}(y)}$ is independent, up to an isomorphism, of $y \in N$. The foliation $\tilde{\mathcal{F}}|_{s^{-1}(y)}$ is a Lie foliation given by a canonical $\mathfrak{g}$-valued Maurer–Cartan form with dense holonomy group.

Proof Part (i) is Theorem 4.20. Part (ii) is an $O(q)$-equivariant application of Corollary 4.25 and Theorem 4.3 (v). The second sentence of part (iii) is a special case of Theorem 4.24. The structure is independent of the choice of the point y, as asserted in the first sentence of (iii), by homogeneity of the $O(q)$-equivariant foliation $\tilde{\mathcal{F}}$ of $OF(M, \mathcal{F})$ and its associated basic foliation (see Theorem 4.3 and Theorem 4.8, and observe that $O(q)$ acts transitively on the components of $OF(M, \mathcal{F})$). □

REMARK. Since $s: OF(M, \mathcal{F}) \to N$ is a fibre bundle, any point y of N has an open neighbourhood U for which there is a diffeomorphism $s^{-1}(U) \cong s^{-1}(y) \times U$ over U. In fact, this diffeomorphism can be chosen so that it maps the foliation $\tilde{\mathcal{F}}|_{s^{-1}(U)}$ to the product of the foliation $\tilde{\mathcal{F}}|_{s^{-1}(y)}$ with the trivial foliation of dimension 0 of U. In other words, the map s is locally trivial even when its fibres are viewed as foliated manifolds. In fact, the local trivialization by the maps T and $\bar{T}$ constructed in the proof of Theorem 4.9 shows that this is the case.

5

Lie groupoids

In Chapter 2, we have discussed the construction of the holonomy group of a given point on a foliated manifold $(M, \mathcal{F})$. This group carries information about the local structure of the foliation near the leaf passing through the given point. Easy examples show that nearby points can have quite different holonomy groups, a phenomenon which typically occurs in the presence of non-compact leaves. In this chapter we shall see that, nevertheless, the holonomy covering spaces of all the leaves of the foliation $(M, \mathcal{F})$ can be fitted together nicely into a smooth (not necessarily Hausdorff) manifold, denoted by $\mathrm{Hol}(M, \mathcal{F})$. Moreover, this manifold carries a partial multiplication operation, which incorporates all the group structures of the various holonomy groups. The resulting structure is that of a Lie groupoid, and the manifold $\mathrm{Hol}(M, \mathcal{F})$ is referred to as the holonomy groupoid of the foliation. This Lie groupoid plays a central role in foliation theory, because it lies at the basis of many constructions of invariants of a foliation, such as the characteristic classes of its normal bundle, the C^*-algebra of the foliation and its K-theory, and the cyclic cohomology of the foliation.

The purpose of this chapter is to provide an introduction to the basic concepts of the theory of Lie groupoids, with special emphasis on the holonomy groupoid of a foliation. In Section 5.1 we introduce Lie groupoids, while in Section 5.2 we give a precise definition of the holonomy groupoid of a foliation and prove some of its first properties. In the next couple of sections, we study various constructions and properties of Lie groupoids, and we discuss the important notion of 'weak equivalence' between Lie groupoids.

Next, in Section 5.5, we introduce the special class of étale groupoids. These étale groupoids are particularly relevant in the context of foliations. On the one hand, they are much easier to handle and much better understood than general Lie groupoids. In fact, they behave in many respects like ordinary manifolds. On the other hand, we shall see that one can obtain an étale groupoid by restricting the holonomy groupoid of a foliation to a 'complete transversal'. Different choices of transversals give weakly equivalent étale groupoids, so that the foliation gives rise to a uniquely determined equivalence class of étale groupoids. This equivalence class is the best model known for the singular 'leaf space' of the foliation.

In Section 5.6, we study proper Lie groupoids, and prove that orbifolds can be seen as Lie groupoids which are both proper and étale. This becomes important when one studies invariants of orbifolds and maps between orbifolds. Finally, with a view to applications in Chapter 6, we close the chapter with a brief discussion of principal bundles, where both the structure group and the base space are replaced by Lie groupoids.

5.1 Definition and first examples

A *groupoid* is a category in which every arrow is invertible. Explicitly, a groupoid G consists of two sets, a set of *objects* G_0 and a set of *arrows* G_1. Each arrow g of G has two objects assigned to it, its source $s(g)$ and its target $t(g)$. We write

$$g\colon x \to y$$

to indicate that $x = s(g)$ and $y = t(g)$. Furthermore, there is an associative multiplication of such arrows for which *source* and *target* match, giving an arrow $hg\colon x \to z$ for any two arrows $g\colon x \to y$ and $h\colon y \to z$. Each arrow $g\colon x \to y$ has an inverse arrow $g^{-1}\colon y \to x$, and for any object x there is a unit $1_x\colon x \to x$.

A well-known example of a groupoid is the fundamental groupoid of a manifold M. The set of objects of this groupoid is M, the arrows from $x \in M$ to $y \in M$ are the homotopy classes of paths (relative to end-points) in M from x to y, and the partial multiplication is induced by the concatenation of paths.

More formally, the structure of a groupoid is given by five *structure maps* relating G_0 and G_1, namely a *source map* $s\colon G_1 \to G_0$ and a *target map* $t\colon G_1 \to G_0$, a *multiplication map* $G_1 \times_{G_0} G_1 \to G_1$, $(h, g) \mapsto hg$, which is defined for those arrows h, g of G with $s(h) = t(g)$, a *unit map* $G_0 \to G_1$, $x \mapsto 1_x$, and an *inverse map* $G_1 \to G_1$, $g \mapsto g^{-1}$. These maps should satisfy the following identities:

(i) $s(hg) = s(g)$, $t(hg) = t(h)$,

(ii) $k(hg) = (kh)g$,

(iii) $1_{t(g)}g = g = g1_{s(g)}$, and

(iv) $s(g^{-1}) = t(g)$, $t(g^{-1}) = s(g)$, $g^{-1}g = 1_{s(g)}$, $gg^{-1} = 1_{t(g)}$

for any $k, h, g \in G_1$ with $s(k) = t(h)$ and $s(h) = t(g)$. The structure maps of a groupoid G together fit into a diagram

$$G_1 \times_{G_0} G_1 \longrightarrow G_1 \longrightarrow G_1 \overset{s}{\underset{t}{\rightrightarrows}} G_0 \longrightarrow G_1 . \qquad (5.1)$$

The conditions (i)–(iv) above can of course be expressed by commutative diagrams involving the structure maps. The *units* of G are the arrows in the image of the unit map of G. Sometimes we say that G is a groupoid *over* the set G_0, and that G_0 is the *base* of the groupoid G. For any $x, y \in G_0$ we write

$$G(x, y) = \{g \in G_1 \mid s(g) = x, \, t(g) = y\} .$$

For any arrow $g \colon x \to y$ (i.e. $g \in G(x, y)$) one says that g is an arrow *from* x *to* y. Furthermore we denote the fibres of s and t by $G(x, \text{-}) = s^{-1}(x)$ and $G(\text{-}, y) = t^{-1}(y)$. The set of arrows from x to x is a group, called the *isotropy group* of G at x, and denoted by

$$G_x = G(x, x) .$$

A *homomorphism between groupoids* H and G is a functor $\phi \colon H \to G$; it is given by a map $H_0 \to G_0$ on objects and a map $H_1 \to G_1$ on arrows, both denoted again by ϕ, which together preserve the groupoid structure, i.e. $\phi(s(h)) = s(\phi(h))$, $\phi(t(h)) = t(\phi(h))$, $\phi(1_y) = 1_{\phi(y)}$ and $\phi(hk) = \phi(h)\phi(k)$ (this implies that also $\phi(h^{-1}) = \phi(h)^{-1}$), for any $h, k \in H$ with $s(h) = t(k)$ and any $y \in H_0$.

A *Lie groupoid* is a groupoid G together with the structure of a smooth (Hausdorff) manifold on the base G_0 and the structure of a (perhaps non-Hausdorff, non-second-countable) smooth manifold on G_1, such that the source map of G is a smooth submersion with Hausdorff fibres and all the other structure maps of G are smooth. Note that the domain of the multiplication map, $G_2 = G_1 \times_{G_0} G_1$, has a natural smooth manifold structure because the source map is a submersion. Also note that it follows that the target map of G is a submersion.

Here, we allowed explicitly that G_1 may be non-Hausdorff and non-second-countable, as this situation arises in our main examples. A Lie groupoid G is called *Hausdorff* if the manifold of arrows G_1 is Hausdorff.

A *homomorphism between Lie groupoids* H and G is by definition a functor $\phi\colon H \to G$ which is smooth, both on objects and on arrows. We say that ϕ is a submersion if $\phi\colon H_1 \to G_1$ is a submersion; this implies that $\phi\colon H_0 \to G_0$ is also a submersion. Lie groupoids and homomorphisms between them form a category.

Examples 5.1 (1) Any manifold M can be viewed as a Lie groupoid over M in which all the arrows are units, i.e. the manifold of arrows is also M. We denote this Lie groupoid again by M, and refer to it as the *unit groupoid* associated to M.

(2) Any manifold M gives rise to another Lie groupoid Pair(M) over M, called the *pair groupoid* of M, with arrows Pair$(M)_1 = M \times M$. The source and the target map are the first and the second projection. The multiplication is unique, because for any $x, x' \in M$ there is exactly one arrow from x to x'. Note that any smooth map $p\colon N \to M$ induces a homomorphism of pair groupoids $p \times p\colon$ Pair$(N) \to$ Pair(M) in the obvious way. Furthermore, if p is a submersion we may define the *kernel groupoid* Ker(p) over N, which is a Lie subgroupoid of Pair(N), consisting of all pairs $(y, y') \in N \times N$ with $p(y) = p(y')$, i.e. Ker$(p)_1 = N \times_M N$.

(3) Any Lie group G can be viewed as a Lie groupoid over a one point space, and with G as the manifold of arrows. We shall denote this Lie groupoid again by G. More generally, any bundle of Lie groups can be viewed as a Lie groupoid for which s and t coincide.

(4) Let M be a manifold. Any immersed submanifold R of $M \times M$ defining an equivalence relation on M such that $\mathrm{pr}_1, \mathrm{pr}_2\colon R \to M$ are submersions gives a Lie groupoid G over M with $G_1 = R$, which is an immersed Lie subgroupoid of the pair groupoid of M. In fact, any Lie groupoid G with the property that $(\mathrm{s}, \mathrm{t})\colon G_1 \to G_0 \times G_0$ is an injective immersion is defined by an equivalence relation in this way.

(5) If G is a Lie group acting smoothly from the left on a manifold M, we define the associated *translation* (or *action*) *groupoid* $G \ltimes M$ over M in which $(G \ltimes M)_1 = G \times M$. The source map is the first projection, the target is given by the action map, and the multiplication is defined by

$$(g', x')(g, x) = (g'g, x).$$

The semi-direct product symbol is used because this construction is a special case of the semi-direct product construction described in Section 5.3.

(6) Let M be a manifold. The *fundamental groupoid* $\Pi(M)$ of M is a Lie groupoid over M in which the arrows from $x \in M$ to $y \in M$ are

the homotopy classes of paths (relative to end-points) in M from x to y, while the multiplication is induced by the concatenation of paths. It is not difficult to show that $\Pi(M)_1$ has indeed a natural smooth structure such that $\Pi(M)$ is a Lie groupoid; we shall not do this here, however, because $\Pi(M)$ is a special case of the monodromy groupoid of a foliation discussed in detail in Section 5.2.

(7) Let E be a vector bundle over a manifold M. One can define a Lie groupoid $GL(E)$ over M such that the arrows from $x \in M$ to $y \in M$ are the linear isomorphisms $E_x \to E_y$ between the fibres of E.

(8) Let G be a Lie group and $\pi\colon P \to M$ a (right) principal G-bundle. Define the *gauge groupoid* Gauge(P) associated to P to be the Lie groupoid over M for which the manifold of arrows is the orbit space of the diagonal action of G on $P \times P$, and the source and the target map are induced by the composition of the first and the second projection with π. Multiplication of arrows in the gauge groupoid is defined in such a way as to make the quotient map $P \times P \to$ Gauge(P) a homomorphism from the pair groupoid.

A *global bisection* of a Lie groupoid G is a section $\sigma\colon G_0 \to G_1$ of s such that $\mathrm{t} \circ \sigma\colon G_0 \to G_0$ is a diffeomorphism. The global bisections of G form a group, which we shall call the *gauge group* of G. The multiplication is given by

$$(\sigma'\sigma)(x) = \sigma'(\mathrm{t}(\sigma(x)))\sigma(x) \, .$$

The unit in this group is the unit map of G.

Example 5.2 Let Gauge(P) be the gauge groupoid associated to a principal G-bundle P over M as in Example 5.1 (8). Since the diagram

$$
\begin{array}{ccc}
P \times P & \longrightarrow & P \times P/G = \mathrm{Gauge}(P)_1 \\
{\scriptstyle \mathrm{pr}_1} \big\downarrow & & \big\downarrow {\scriptstyle \mathrm{s}} \\
P & \longrightarrow & M
\end{array}
$$

is a pull-back, any section σ of s induces a section $\tilde{\sigma} = (\mathrm{id}, \bar{\sigma})\colon P \to P \times P$ of pr_1. Here $\bar{\sigma}\colon P \to P$ is a G-equivariant diffeomorphism. Conversely, any such a G-equivariant diffeomorphism induces a section of s. The gauge group of Gauge(P) may be therefore identified with the group of G-equivariant diffeomorphisms $P \to P$.

A *local bisection* of a Lie groupoid G is a local section $\sigma\colon U \to G$ of s, defined on an open subset U of G_0, such that $\mathrm{t} \circ \sigma$ is an open embedding.

The germs of bisections of G form a groupoid, which we shall denote by $\mathrm{Bis}(G)$. The multiplication in $\mathrm{Bis}(G)$ is locally given by the same formula as the multiplication in the gauge group of G above. With the standard sheaf topology $\mathrm{Bis}(G)$ becomes a Lie groupoid over G_0, with $\dim \mathrm{Bis}(G)_1 = \dim G_0$ (a Lie groupoid with this property is called *étale*; see Section 5.5). The obvious functor $\mathrm{Bis}(G) \to G$ is smooth; it is also surjective, as the following proposition shows.

Proposition 5.3 *Let G be a Lie groupoid. For any $g \in G$ there exists a local bisection $\sigma \colon U \to G$ of G with $g \in \sigma(U)$.*

Proof Choose a subspace V of $T_g(G_1)$ which is complementary to both $\mathrm{Ker}(ds)_g$ and $\mathrm{Ker}(dt)_g$. Choose a local section $\sigma \colon U \to G$ (e.g. in local coordinates) of s such that $(d\sigma)_{\mathrm{s}(g)}(T_{\mathrm{s}(g)}(G_0)) = V$. It follows that $d(\mathrm{t} \circ \sigma)_{\mathrm{s}(g)}$ is an isomorphism, so we can shrink U if necessary so that $\mathrm{t} \circ \sigma$ becomes an open embedding. $\qquad\square$

Let G be a Lie groupoid, and $x \in G_0$. The source and the target map of G are submersions, therefore the fibres $G(x, \text{-}) = \mathrm{s}^{-1}(x)$ and $G(\text{-}, x) = \mathrm{t}^{-1}(x)$ are closed submanifolds of G_1. Note that the right action of the isotropy group G_x on $G(x, \text{-})$ is free and transitive along the fibres of $\mathrm{t}_x = \mathrm{t}|_{G(x,\text{-})}$. The *orbit* of G passing through x is by definition

$$Gx = \mathrm{t}(G(x, \text{-})) \subset G_0 \ .$$

The following theorem will in particular show that the isotropy groups are in fact Lie groups, and that the orbits are immersed submanifolds of G_0.

Theorem 5.4 *Let G be a Lie groupoid, and let $x, y \in G_0$.*

(i) $G(x, y)$ is a closed submanifold of G.

(ii) G_x is a Lie group.

(iii) Gx is an immersed submanifold of G_0.

(iv) $\mathrm{t}_x \colon G(x, \text{-}) \to Gx$ is a principal G_x-bundle.

REMARK. Since G_x is a Lie group, we may consider the connected component G_x° of the unit in G_x. Theorem 5.4 (iv) implies that $G(x, \text{-})/G_x^{\circ}$ is a Hausdorff manifold, and that $\mathrm{t}_x \colon G(x, \text{-}) \to Gx$ factors through the quotient projection $G(x, \text{-}) \to G(x, \text{-})/G_x^{\circ}$ as a covering projection

$G(x, \text{-})/G_x^{\circ} \to Gx$. The group of covering transformations of this covering projection is the discrete group of components $\pi_0(G_x) = G_x/G_x^{\circ}$ of G_x.

In the proof this theorem we shall use the following result, which describes when the quotient of a free (not necessarily compact) Lie group action has a smooth (perhaps non-Hausdorff) structure.

Lemma 5.5 *Let $\mu\colon M \times G \to M$ be a free action of a Lie group G on a manifold M. The following conditions are equivalent.*

(i) For any $x \in M$ there exists an embedded submanifold U with $x \in U$ such that $U \times G \to M$ given by the action of G is an open embedding.

(ii) There is a smooth (perhaps non-Hausdorff) structure on M/G such that the quotient projection $M \to M/G$ is a principal G-bundle.

(iii) There exist a (perhaps non-Hausdorff) manifold N and a smooth map $f\colon M \to N$ which is constant on the G-orbits and satisfies

$$\mathrm{Ker}(df)_x = (d\mu)_{(x,1)}(\{0\} \times T_1(G)) .$$

REMARK. Recall that the connected components of orbits of a free Lie group action give us a foliation of the manifold. If the Lie group is compact, the slice theorem (or the Reeb stability theorem, Theorem 2.9) shows that (i) above is fulfilled.

Proof (of Lemma 5.5) If (i) holds, then $U \to M/G$ is a topological open embedding, and we may define a smooth structure on M/G such that this map is a smooth open embedding, for any such U. Therefore (i) implies (ii). Note that (iii) follows directly from (ii). So we only need to prove that (iii) implies (i).

Take any $x \in M$, and choose a submersion $h\colon V \to \mathbb{R}^k$ defined on an open neighbourhood V of $f(x)$ in N such that $\mathrm{Ker}(dh)_{f(x)}$ is complementary to $(df)_x(T_x(M))$. Next, choose a small transversal section U of the foliation of M given by the connected components of the G-orbits, with $x \in U$ and $f(U) \subset V$. Now, by construction, $(d(h \circ f|_U))_x$ is an isomorphism, so we may shrink U if necessary so that

$$h \circ f|_U$$

is an open embedding. In particular, f is injective on U. Since f is also constant along the G-orbits, it follows that each G-orbit intersects U in at most one point. Since U is transversal to the G-orbits, this proves (i). $\qquad\square$

We shall now give a proof of Theorem 5.4.

Proof (of Theorem 5.4) Put $E_g = \text{Ker}(ds)_g \cap \text{Ker}(dt)_g$, for any $g \in G$. We will show that $E|_{G(x,\text{-})}$ is an involutive subbundle of the tangent bundle of $G(x,\text{-})$ and hence defines a foliation $\mathcal{F}_x$ of $G(x,\text{-})$.

For any $g \in G(x,\text{-})$, the left translation by g gives us a diffeomorphism

$$L_g \colon G(\text{-},x) \longrightarrow G(\text{-},t(g))$$

by $L_g(h) = gh$. For any $h \in G(\text{-},x)$ note that E_h is a subspace of $T_h(G(\text{-},x)) \subset T_h(G)$. Since $s \circ L_g = s|_{G(\text{-},x)}$, it follows that

$$(dL_g)_{1_x}(E_{1_x}) = E_g \ .$$

Furthermore, any basis $v_1, \dots, v_k$ of E_{1_x} can be extended to a global frame $X_1, \dots, X_k$ of $E|_{G(x,\text{-})}$ by $(X_i)_g = (dL_g)_{1_x}(v_i)$.

This shows that $E|_{G(x,\text{-})}$ is indeed a subbundle of the tangent bundle of $G(x,\text{-})$. It is involutive because it is exactly the kernel of the derivative of the map t_x. Hence it defines a foliation $\mathcal{F}_x$ of $G(x,\text{-})$ (which is parallelizable by the frame $X_1, \dots, X_k$). The leaves of $\mathcal{F}_x$ are exactly the connected components of the fibres of t_x. So these fibres are closed manifolds, proving (i). In particular, the fibre $t_x^{-1}(x) = G_x$ is a Lie group.

The Lie group G_x acts smoothly and freely on $G(x,\text{-})$ from the right, and transitively along the fibres of t_x. Note that the condition (iii) of Lemma 5.5 is fulfilled by the map t_x, so the proposition implies that there is a natural structure of a smooth manifold on the orbit Gx making $t_x \colon G(x,\text{-}) \to Gx$ into a principal G_x-bundle. The fact that G_0 is Hausdorff implies that Gx is also Hausdorff. $\qquad\qquad\Box$

5.2 The monodromy and holonomy groupoids

In this section we will discuss the important construction of the holonomy groupoid of a foliation, as well as that of the related monodromy groupoid. These Lie groupoids play a central role in many of the constructions of invariants of foliations.

Throughout this section, $(M, \mathcal{F})$ denotes a fixed foliated manifold. Recall that each leaf L of $\mathcal{F}$ has a natural smooth structure for which the inclusion $L \to M$ is an immersion (Section 1.1). We now describe the two groupoids, and prove that they are indeed Lie groupoids.

First, the *monodromy groupoid* $\text{Mon}(M, \mathcal{F})$ of $(M, \mathcal{F})$ is a groupoid over M with the following arrows:

(a) if $x, y \in M$ lie on the same leaf L of $\mathcal{F}$, then the arrows in $\mathrm{Mon}(M, \mathcal{F})$ from x to y are the homotopy classes (relative to end-points) of paths in L from x to y, while

(b) if $x, y \in M$ lie on different leaves of $\mathcal{F}$, there are no arrows between them.

The multiplication is induced by the concatenation of paths. In particular, the isotropy groups of the monodromy groupoid are the fundamental groups of the leaves, i.e.

$$\mathrm{Mon}(M, \mathcal{F})_x = \pi_1(L, x)$$

for any point x on a leaf L.

Now the *holonomy groupoid* $\mathrm{Hol}(M, \mathcal{F})$ is defined analogously, except that one takes the holonomy classes of paths for arrows instead of the homotopy classes. Here we have

$$\mathrm{Hol}(M, \mathcal{F})_x = \mathrm{Hol}(L, x)$$

for any point x on a leaf L (Section 2.1). Note that, by property (ii) of holonomy from Section 2.1 (page 22), there is a natural quotient homomorphism $\mathrm{Mon}(M, \mathcal{F}) \to \mathrm{Hol}(M, \mathcal{F})$.

Proposition 5.6 *The monodromy groupoid of a foliation has a natural Lie groupoid structure. The same is true for the holonomy groupoid.*

Proof We shall prove this for the monodromy groupoid; for the holonomy groupoid the proof is analogous. We shall define a base for topology on $\mathrm{Mon}(M, \mathcal{F})_1$ which consists of pairwise compatible local charts. Take any path σ from x to y in a leaf L of $\mathcal{F}$, and denote by ς the associated arrow in $\mathrm{Mon}(M, \mathcal{F})$. Choose a foliation chart $\varphi \colon U \to \mathbb{R}^p \times \mathbb{R}^q$ with $x \in U$, and a foliation chart $\psi \colon V \to \mathbb{R}^p \times \mathbb{R}^q$ with $y \in V$. We can assume that the image of φ is of the form $A \times C$ and that the image of ψ is of the form $B \times D$, where A and B are connected and simply connected open subsets of $\mathbb{R}^p$, while C and D are connected and simply connected open subsets $\mathbb{R}^q$. Write $\varphi(x) = (a, c)$ and $\psi(y) = (b, d)$. We have the transversal section $S = \varphi^{-1}(\{a\} \times C)$ at x and the transversal section $T = \psi^{-1}(\{b\} \times D)$ at y. We may also assume that these two sections are so small that the holonomy homomorphism $\mathrm{hol}^{T,S}(\sigma) \colon (S, x) \to (T, y)$ is a diffeomorphism from S onto T. This means that there is a smooth map $H \colon [0, 1] \times S \to M$ such that $H(\,\text{-}\,, z)$ is a path in a leaf of $\mathcal{F}$, $H(0, z) = z$ and $H(1, z) = \mathrm{hol}^{T,S}(\sigma)(z)$ for any $z \in S$. Now define an injective map

$$f\colon A \times B \times C \longrightarrow \mathrm{Mon}(M, \mathcal{F})_1$$

as follows: for any $(a', b', c') \in A \times B \times C$, let $f(a', b', c')$ be the concatenation of paths

$$\tau H(-, z)\gamma \,,$$

where γ is any path from $\varphi^{-1}(a', c')$ to $z = \varphi^{-1}(a, c')$ in the plaque $\varphi^{-1}(A \times \{c'\})$, and τ is any path from $z' = \mathrm{hol}^{T,S}(\sigma)(z)$ (write $\psi(z') = (b, d')$) to $\psi^{-1}(b', d')$ in the plaque $\psi^{-1}(B \times \{d'\})$. Now the image of f is parametrized by f, and we take it for a basic open set around ς. One can check that all such basic open sets form a basis for a topology, and furthermore that they form a smooth atlas of $\mathrm{Mon}(M, \mathcal{F})_1$ which makes it into a Lie groupoid of dimension $2p + q = n - q$. $\square$

We conclude this section with the following observations, which are all immediate from the construction.

Proposition 5.7 *Let $\mathcal{F}$ be a foliation of a manifold M.*

(i) The orbits of the monodromy and the holonomy groupoids of $(M, \mathcal{F})$ (with the smooth structure as in Theorem 5.4) are exactly the leaves of $\mathcal{F}$ (with the smooth structure as in Section 1.1).

(ii) The isotropy groups of the monodromy and the holonomy groupoid of $(M, \mathcal{F})$ are discrete.

(iii) For a point x on a leaf L, the target map of the monodromy groupoid restricts to the universal covering map

$$\mathrm{Mon}(M, \mathcal{F})(x, -) \longrightarrow L \,,$$

while the restriction of the target map of the holonomy groupoid

$$\mathrm{Hol}(M, \mathcal{F})(x, -) \longrightarrow L$$

is the covering projection corresponding to the kernel of the holonomy homomorphism $\pi_1(L, x) \to \mathrm{Hol}(L, x)$.

(iv) The natural quotient map

$$\mathrm{Mon}(M, \mathcal{F}) \longrightarrow \mathrm{Hol}(M, \mathcal{F})$$

is a homomorphism of Lie groupoids, and restricts to a covering projection

$$\mathrm{Mon}(M, \mathcal{F})(x, -) \longrightarrow \mathrm{Hol}(M, \mathcal{F})(x, -)$$

for any $x \in M$.

The holonomy groupoid of a foliation can be difficult to describe in concrete examples. In fact, there are many cases of foliations with non-Hausdorff holonomy groupoid.

Examples 5.8 (1) Let $f\colon M \to N$ be a surjective submersion with connected fibres, and $\mathcal{F}$ the associated foliation of M. Then all the leaves have trivial holonomy, and the holonomy groupoid of $(M, \mathcal{F})$ is $\mathrm{Ker}(f) = M \times_N M$. If the fibres of f are simply connected, then this Lie groupoid is also the monodromy groupoid of $(M, \mathcal{F})$.

(2) Let $\mathcal{F}$ be a foliation of M, invariant under a free properly discontinuous action of a discrete group G such that M/G is a Hausdorff manifold. In particular, we have the foliation $\mathcal{F}/G$ on $M/\mathcal{F}$. Now let $\mathrm{Mon}(M, \mathcal{F})$ be the monodromy groupoid of $(M, \mathcal{F})$. Note that the action of G induces an action on the paths in M and also on $\mathrm{Mon}(M, \mathcal{F})$ because the action respects the foliation. Furthermore this action is again free and properly discontinuous, and we get a Lie groupoid

$$\mathrm{Mon}(M, \mathcal{F})/G$$

over M/G with $(\mathrm{Mon}(M, \mathcal{F})/G)_1 = \mathrm{Mon}(M, \mathcal{F})_1/G$. Since we can identify paths in the leaves of $\mathcal{F}/G$ with the equivalence classes of paths in the leaves of $\mathcal{F}$, and because this identification also respects the homotopy, we get

$$\mathrm{Mon}(M, \mathcal{F})/G \cong \mathrm{Mon}(M/G, \mathcal{F}/G) \, .$$

A similar construction gives us a Lie groupoid $\mathrm{Hol}(M, \mathcal{F})/G$ and a surjective homomorphism

$$\mathrm{Hol}(M, \mathcal{F})/G \longrightarrow \mathrm{Hol}(M/G, \mathcal{F}/G) \, ,$$

which however is not an isomorphism in general.

(3) Let G be a discrete group acting freely and properly discontinuously (from the right) on a connected manifold $\tilde{M}$ such that $\tilde{M}/G = M$ is a Hausdorff manifold. Suppose that we also have a left action of G on a manifold F. Now consider the associated foliation $\mathcal{F}$ on the flat bundle $\tilde{M} \times_G F$ (suspension) over M. By definition $\mathcal{F} = \mathcal{G}/G$, where $\mathcal{G}$ is the foliation on $\tilde{M} \times F$ given by the second projection. By (2) we have

$$\mathrm{Mon}(M, \mathcal{G})/G \cong \mathrm{Mon}(\tilde{M} \times_G F, \mathcal{F}) \, ,$$

which equals $(\tilde{M} \times \tilde{M} \times F)/G$ if $\tilde{M}$ is simply connected (by (1)). Also we have a surjective homomorphism

$$\mathrm{Hol}(M, \mathcal{G})/G \longrightarrow \mathrm{Hol}(\tilde{M} \times_G F, \mathcal{F}) \ .$$

(4) Let $\mathcal{F}$ be the standard foliation of the Möbius band M (see Example 1.1 (4)). This is a special case of a suspension, so the monodromy groupoid can be computed by (3) as

$$\mathrm{Mon}(M, \mathcal{F}) = (\mathbb{R} \times \mathbb{R} \times (-1, 1))/\mathbb{Z} \ .$$

All the source-fibres of the holonomy groupoid, however, are diffeomorphic to S^1.

(5) The Kronecker foliation $\mathcal{F}$ of torus T^2 (see Example 1.1 (3)) is also a suspension, but here the holonomy and the monodromy groupoids coincide,

$$\mathrm{Hol}(T^2, \mathcal{F}) = \mathrm{Mon}(T^2, \mathcal{F}) = (\mathbb{R} \times \mathbb{R} \times S^1)/\mathbb{Z} \ .$$

(6) Let $\mathcal{F}$ be the Reeb foliation of S^3. The compact leaf of $\mathcal{F}$ has $\mathbb{R}^2$ for the holonomy cover, while any other leaf is itself diffeomorphic to $\mathbb{R}^2$ and has trivial holonomy group. Since the fibre of the source map is the holonomy cover of the corresponding leaf, the holonomy groupoid is the same as the monodromy groupoid and as a set it is the product $S^3 \times \mathbb{R}^2$. However, the topology of this space is not the product topology. In fact, one can see that this Lie groupoid is not Hausdorff.

(7) Let $\mathcal{F}$ be a Riemannian foliation on M. Then the derivative $\mathrm{Hol}(M, \mathcal{F}) \to GL(N(\mathcal{F}))$ is an injective homomorphism. In particular, the holonomy groupoid of a Riemannian foliation is Hausdorff.

(8) The monodromy groupoid of a transversely orientable foliation of codimension 1 is Hausdorff if and only if the foliation has no vanishing cycles (as defined in Subsection 3.2.1). The same observation holds true also for any foliation, if one extends the notion of a vanishing cycle to arbitrary foliations in the obvious way.

5.3 Some general constructions

In this section we will define transformations between homomorphisms of Lie groupoids, and discuss some ways of constructing new Lie groupoids out of given ones.

Induced groupoids. Let G be a Lie groupoid and $\phi \colon M \to G_0$ a smooth map. Then one can define the *induced groupoid* $\phi^*(G)$ over M

in which the arrows from x to y are the arrows in G from $\phi(x)$ to $\phi(y)$, i.e.

$$\phi^*(G)_1 = M \times_{G_0} G_1 \times_{G_0} M \,,$$

and the multiplication is given by the multiplication in G. The space $\phi^*(G)_1$ can be constructed by two pull-backs as in the diagram

$$
\begin{array}{ccc}
\phi^*(G)_1 & \longrightarrow & M \\
\downarrow & & \downarrow {\scriptstyle \phi} \\
G_1 \times_{G_0} M \xrightarrow{\ \mathrm{pr}_1\ } G_1 & \xrightarrow{\ t\ } & G_0 \\
\downarrow & \downarrow {\scriptstyle s} & \\
M & \xrightarrow{\ \phi\ } & G_0
\end{array}
$$

The lower pull-back has a natural smooth structure because s is a submersion. If the composition $t \circ \mathrm{pr}_1$ is also a submersion, the upper pull-back has a natural smooth structure as well. It follows that the diagram

$$
\begin{array}{ccc}
\phi^*(G)_1 & \longrightarrow & G_1 \\
{\scriptstyle (s,t)} \downarrow & & \downarrow {\scriptstyle (s,t)} \\
M \times M & \xrightarrow{\ \phi \times \phi\ } & G_0 \times G_0
\end{array}
$$

is a pull-back. Therefore $\phi^*(G)$ is a Lie groupoid whenever the map

$$t \circ \mathrm{pr}_1 \colon G_1 \times_{G_0} M \longrightarrow G_0$$

is a submersion. The map ϕ induces a homomorphism of Lie groupoids $\phi \colon \phi^*(G) \to G$.

Transformations. For two homomorphisms $\phi, \psi \colon G \to H$ of Lie groupoids, a (smooth) *natural transformation* (briefly transformation) from ϕ to ψ is a smooth map

$$T \colon G_0 \longrightarrow H_1$$

such that for each $x \in G_0$, $T(x)$ is an arrow from $\phi(x)$ to $\psi(x)$ in H, and for each arrow $g \colon x \to y$ in G the square

$$
\begin{array}{ccc}
\phi(x) & \xrightarrow{\ T(x)\ } & \psi(x) \\
{\scriptstyle \phi(g)} \downarrow & & \downarrow {\scriptstyle \psi(g)} \\
\phi(y) & \xrightarrow{\ T(y)\ } & \psi(y)
\end{array}
$$

commutes. We write $T\colon \phi \to \psi$ to indicate that T is such a transformation from ϕ to ψ.

If $T\colon \phi \to \psi$ and $R\colon \psi \to \rho$ are two transformations, so is their product $RT\colon \phi \to \rho$ given by $RT(x) = R(x)T(x)$. In particular, the homomorphisms from G to H are themselves the objects of a groupoid with transformations as arrows. In fact, Lie groupoids, homomorphisms and transformations form a 2-category.

Sums and products. For two Lie groupoids G and H one can construct the *product*

$$G \times H$$

in the obvious way, by taking the product manifolds $G_0 \times H_0$ and $G_1 \times H_1$. In a similar way one constructs the *sum* (disjoint union)

$$G + H \ .$$

The sums and products have familiar universal properties in the category (in fact, also in the 2-category) of Lie groupoids and homomorphisms.

Strong fibred products. For two homomorphisms $\phi\colon G \to K$ and $\psi\colon H \to K$ one can construct the fibred products of the sets of objects and arrows:

$$(G \times_K H)_0 = G_0 \times_{K_0} H_0 = \{(x,y) \mid x \in G_0,\, y \in H_0,\, \phi(x) = \psi(y)\} \ ,$$

$$(G \times_K H)_1 = G_1 \times_{K_1} H_1 = \{(g,h) \mid g \in G_1,\, h \in H_1,\, \phi(g) = \psi(h)\} \ .$$

With multiplication defined component-wise, this defines a groupoid

$$G \times_K H \ .$$

However, in general this is not a Lie groupoid. It is, if the fibred products $G_0 \times_{K_0} H_0$ and $G_1 \times_{K_1} H_1$ are transversal. For example, for $G_0 \times_{K_0} H_0$ this means that the map $\phi \times \psi\colon G_0 \times H_0 \to K_0 \times K_0$ is transversal to the diagonal $\Delta\colon K_0 \to K_0 \times K_0$, so that $G_0 \times_{K_0} H_0 = (\phi \times \psi)^{-1}(\Delta K_0)$ is indeed a manifold.

If the transversality condition is satisfied, this construction gives a (strong) *fibred product* (pull-back) with the familiar universal property. Below we will consider an alternative, larger fibred product. To emphasize the distinction, we often refer to the present fibred product as the *strong* one.

Weak fibred products. Let $\phi\colon G \to K$ and $\psi\colon H \to K$ be homomorphisms of Lie groupoids. We define a new groupoid P as follows. Objects of P are triples (x, k, y), where $x \in G_0$, $y \in H_0$ and $k \in K(\phi(x), \psi(y))$. Arrows in P from (x, k, y) to (x', k', y') are pairs (g, h) of arrows $g \in G_1$ and $h \in H_1$ such that

$$k'\phi(g) = \psi(h)k .$$

The multiplication is given component-wise. Often (but not always) P has the structure of a Lie groupoid. Indeed, the set of objects may be considered as the fibred product

$$P_0 = G_0 \times_{K_0} K_1 \times_{K_0} H_0 ,$$

and if the fibred product is transversal then P_0 inherits a natural structure of a submanifold of $G_0 \times K_1 \times H_0$. This is the case, for example, when either $\phi\colon G_0 \to K_0$ or $\psi\colon H_0 \to K_0$ is a submersion. If P_0 has a manifold structure as above, then

$$P_1 = G_1 \times_{K_0} K_1 \times_{K_0} H_1 = \{(g, k, h) \mid \phi(\mathrm{s}(g)) = \mathrm{s}(k),\ \psi(\mathrm{s}(h)) = \mathrm{t}(k)\}$$

is also a manifold. Indeed, in this case P_1 can be obtained from the two fibred products

$$
\begin{array}{ccc}
G_1 \times_{K_0} K_1 \times_{K_0} H_1 & \longrightarrow & H_1 \\
\downarrow & & \downarrow{\scriptstyle \mathrm{s}} \\
G_1 \times_{K_0} K_1 \times_{K_0} H_0 \longrightarrow G_0 \times_{K_0} K_1 \times_{K_0} H_0 \xrightarrow{\ \mathrm{pr}_3\ } & H_0 \\
\downarrow & \downarrow{\scriptstyle \mathrm{pr}_1} & \\
G_1 \xrightarrow{\qquad \mathrm{s} \qquad} & G_0 &
\end{array}
$$

In this case P is a Lie groupoid, referred as the *weak pull-back* or the *weak fibred product*, and denoted by

$$G \times_K^{(\mathrm{w})} H .$$

REMARK. We will use weak fibred products more often than strong ones, and if not stated explicitly otherwise, 'fibred product' from now on will refer to the weak one, and will be simply denoted by $G \times_K H$.

For completeness, we now discuss the universal property of weak fibred products, but we shall not make explicit use of it. Let P be the weak fibred product as above, and consider the following square of homomorphisms of Lie groupoids.

$$
\begin{array}{ccc}
P & \xrightarrow{\ \mathrm{pr}_3\ } & H \\
{\scriptstyle \mathrm{pr}_1}\big\downarrow & & \big\downarrow{\scriptstyle \psi} \\
G & \xrightarrow{\ \phi\ } & K
\end{array}
$$

This square does not commute, but there is an obvious transformation $T\colon \phi \circ \mathrm{pr}_1 \to \psi \circ \mathrm{pr}_3$, defined by

$$T(x, k, y) = k \ .$$

For any Lie groupoid Q, let $\mathrm{Hom}(Q, \phi, \psi)$ be the groupoid with objects the triples (ρ_1, ρ_2, R), where $\rho_1 \colon Q \to G$ and $\rho_2 \colon Q \to H$ are homomorphisms and $R \colon \phi \circ \rho_1 \to \psi \circ \rho_2$ is a transformation. The arrows $(\rho_1, \rho_2, R) \to (\sigma_1, \sigma_2, S)$ in $\mathrm{Hom}(Q, \phi, \psi)$ are pairs of transformations $U_i \colon \rho_i \to \sigma_i$ $(i = 1, 2)$ such that $\psi(U_2)R = S\phi(U_1)$. Then there is a functor

$$\mathrm{Hom}(Q, P) \longrightarrow \mathrm{Hom}(Q, \phi, \psi)$$

defined by composition with the square above. On objects, this functor sends $\alpha \colon Q \to P$ to

$$(\mathrm{pr}_1 \circ \alpha, \mathrm{pr}_3 \circ \alpha, T\alpha) \ .$$

The universal property of the weak fibred product can be expressed by stating that this functor is an equivalence of groupoids (for each groupoid Q, and natural in Q).

Semi-direct products. Let G be a Lie groupoid.

(i) A left *action* of G on a manifold N *along* a smooth map $\epsilon \colon N \to G_0$ is given by a smooth map $\mu \colon G_1 \times_{G_0} N \to N$ (we write $\mu(g, y) = gy$), defined on the pull-back $G_1 \times_{G_0} N = \{(g, y) \,|\, s(g) = \epsilon(y)\}$, which satisfies the following identities: $\epsilon(gy) = t(g)$, $1_{\epsilon(y)}y = y$ and $g'(gy) = (g'g)y$, for any $g', g \in G_1$ and $y \in N$ with $s(g') = t(g)$ and $s(g) = \epsilon(y)$. For such an action one can form the *translation groupoid*

$$G \ltimes N$$

over N with $(G \ltimes N)_1 = G_1 \times_{G_0} N$, exactly as for Lie group actions (Example 5.1 (5)). This groupoid is a Lie groupoid, also referred to as the *semi-direct product groupoid* of the G-action.

We define the quotient $G \backslash N$ as the space of orbits of the Lie groupoid $G \ltimes N$. This space is in general not a manifold.

A right action of G on N is defined analogously, and such an action gives a semi-direct product $N \rtimes G$ and a space of orbits N/G.

(ii) There is also a notion of a (right) *action* of a Lie groupoid G on another Lie groupoid H. It is given by two (right) actions of G on H_1 and on H_0, such that the groupoid structure maps of H are equivariant, i.e. compatible with the actions by G. If we denote the action maps on H_i by $\epsilon_i \colon H_i \to G_0$ and $\mu_i \colon H_i \times_{G_0} G_1 \to H_i$, $i = 0, 1$, this implies in particular that $\epsilon_0 \circ s = \epsilon_1 = \epsilon_0 \circ t$, and that the diagonal action of G on the domain $H_1 \times_{H_0} H_1$ of the multiplication map is well-defined. Note that for each $x \in G_0$ the fibre $H_x = \epsilon_1^{-1}(x)$ is a full subgroupoid of H over $\epsilon_0^{-1}(x)$, so that one has a *family* of groupoids indexed by the points $x \in G_0$. The groupoid G 'acts' on this family by a groupoid isomorphism $H_{x'} \to H_x$, for any arrow $g \colon x \to x'$. Note that these H_x are Lie groupoids if ϵ_0 is a submersion, and the action isomorphisms $H_{x'} \to H_x$ are isomorphisms of Lie groupoids.

For such an action of G on H, one can form the *semi-direct product groupoid*

$$H \rtimes G \,,$$

which is a Lie groupoid over H_0. The manifold of arrows of this Lie groupoid is the pull-back

$$(H_0 \times_{G_0} G_1) \times_{H_0} H_1 = \{(y, g, h) \,|\, \epsilon_0(y) = t(g),\, yg = t(h)\} \,.$$

A triple $(y, g, h) \in (H_0 \times_{G_0} G_1) \times_{H_0} H_1$ is an arrow from $s(h)$ to y. These arrows multiply by the usual formula

$$(y, g, h)(y', g', h') = (y, gg', (hg')h') \,.$$

If H is the unit groupoid of a manifold N, this definition agrees with the one given in (i) up to the obvious isomorphism.

Lemma 5.9 *Consider a right action of a Lie groupoid G on a Lie groupoid H. If H_0 is a principal G-bundle over B, then H_1/G is a Lie groupoid over $B \cong H_0/G$.*

Proof The only thing that has to be shown is that H_1/G is a manifold. We can specify the manifold structure locally in B, so it suffices to consider the case where $\pi \colon H_0 \to B$ has a section σ. But then H_1/G is isomorphic to the pull-back of s$\colon H_1 \to H_0$ along $\sigma \colon B \to H_0$, hence is a manifold. Moreover, this manifold structure is independent of the choice of σ, since by principality of the action on H_0, any two sections

σ and σ' are related by a map $\theta \colon B \to G_1$, as $\sigma(b)\theta(b) = \sigma'(b)$ for all $b \in B$. Then the same multiplication by θ establishes a diffeomorphism between the pull-back of s along σ and the one along σ'. $\qquad\square$

REMARK. We denote the Lie groupoid H_1/G over B by H/G. The quotient morphism $H \to H/G$ induces for each $x \in H_0$ an isomorphism of s-fibres $s^{-1}(x) \to s^{-1}(\pi(x))$. More precisely, the square

$$
\begin{array}{ccc}
H_1 & \longrightarrow & H_1/G \\
{\scriptstyle s}\downarrow & & \downarrow{\scriptstyle s} \\
H_0 & \xrightarrow{\;\pi\;} & B = H_0/G
\end{array}
$$

is a pull-back of smooth manifolds.

5.4 Equivalence of Lie groupoids

In this section we discuss some notions of isomorphism and equivalence of Lie groupoids.

Isomorphisms. Two Lie groupoids G and H are said to be *isomorphic* if there are homomorphisms $\phi \colon G \to H$ and $\psi \colon H \to G$ such that $\phi \circ \psi$ and $\psi \circ \phi$ are the identity homomorphisms of H and G respectively. In this case ϕ and ψ are called *isomorphisms*. This terminology agrees with the usual one referring to the category of Lie groupoids and homomorphisms.

Equivalences of categories. Recall that two categories $\mathcal{C}$ and $\mathcal{D}$ (no topology or smooth structure involved) are said to be *equivalent* if there are functors $F \colon \mathcal{C} \to \mathcal{D}$ and $G \colon \mathcal{D} \to \mathcal{C}$, and natural isomorphisms $\tau \colon F \circ G \to \mathrm{id}_\mathcal{D}$ and $\sigma \colon G \circ F \to \mathrm{id}_\mathcal{C}$. Using the axiom of choice, this notion of equivalence can alternatively be described as follows. Two categories $\mathcal{C}$ and $\mathcal{D}$ are equivalent if there is a functor $F \colon \mathcal{C} \to \mathcal{D}$ with the following two properties:

(i) F is *essentially surjective*; that is, for any object y of $\mathcal{D}$ there are an object x of $\mathcal{C}$ and an isomorphism $F(x) \to y$ in $\mathcal{D}$; and

(ii) F is *full and faithful*; that is, for any two objects x and x' in $\mathcal{C}$ the functor F induces a bijection

$$
F \colon \mathcal{C}(x, x') \longrightarrow \mathcal{D}(F(x), F(x'))
$$

between the set of all arrows from x to x' in $\mathcal{C}$ and the set of all arrows from $F(x)$ to $F(x')$ in $\mathcal{D}$.

These two ways of describing equivalence of course apply also to groupoids. However, when some additional structure is involved the two ways are no longer equivalent, and for Lie groupoids we distinguish the two notions described below.

Strong equivalence of Lie groupoids. Let G and H be Lie groupoids. A homomorphism $\phi\colon G \to H$ is called a *strong equivalence* if there are a homomorphism $\psi\colon H \to G$ and transformations $T\colon \phi \circ \psi \to \mathrm{id}_H$ and $S\colon \psi \circ \phi \to \mathrm{id}_G$.

Weak equivalences of Lie groupoids. Let G and H be Lie groupoids. A homomorphism $\phi\colon G \to H$ is called a *weak equivalence* if it satisfies the following two modified conditions for being essentially surjective, and full and faithful:

(ES) the map $\mathrm{t} \circ \mathrm{pr}_1\colon H_1 \times_{H_0} G_0 \to H_0$, sending a pair (h, x) with $\mathrm{s}(h) = \phi(x)$ to $\mathrm{t}(h)$, is a surjective submersion, and

(FF) the square

$$
\begin{array}{ccc}
G_1 & \xrightarrow{\ \phi\ } & H_1 \\
{\scriptstyle (\mathrm{s},\mathrm{t})}\big\downarrow & & \big\downarrow{\scriptstyle (\mathrm{s},\mathrm{t})} \\
G_0 \times G_0 & \xrightarrow{\ \phi \times \phi\ } & H_0 \times H_0
\end{array}
$$

is a fibred product of manifolds.

Strong equivalences are rare, but there are many examples of weak equivalences.

Examples 5.10 (1) Let M be a manifold and $\mathrm{Pair}(M)$ its pair groupoid. The homomorphism $\mathrm{Pair}(M) \to 1$, to the trivial one point groupoid consisting of one object and one arrow, is a strong and a weak equivalence.

(2) Let $p\colon N \to M$ be a surjective submersion between manifolds. We view M as the unit groupoid (Example 5.1 (1)), and consider the kernel groupoid $\mathrm{Ker}(p)$ (Example 5.1 (2)),

$$
N \times_M N \rightrightarrows N \,.
$$

The map p induces a weak equivalence $\mathrm{Ker}(p) \to M$. A particular case of this is where $N = \coprod_i U_i$ is the disjoint union of an open cover $\{U_i\}$

of M, and p is the evident map. Then $\mathrm{Ker}(p)$ takes the form

$$\coprod_{i,j} U_i \cap U_j \rightrightarrows \coprod_i U_i \, .$$

(3) A Lie groupoid G is said to be *transitive* if the map

$$(\mathrm{s},\mathrm{t})\colon G_1 \longrightarrow G_0 \times G_0$$

is a surjective submersion. For any object x of a transitive Lie groupoid G, the inclusion

$$G_x \longrightarrow G$$

of the isotropy group at x into G is a weak equivalence. As a special case, this yields a weak equivalence between the gauge groupoid of a principal G-bundle (Example 5.1 (8)) and the Lie group G.

(4) Suppose that $\phi\colon M \to G_0$ is a smooth map with the property that $\mathrm{t}\circ\mathrm{pr}_1\colon G_1 \times_{G_0} M \to G_0$ is a submersion. Then the homomorphism $\phi\colon \phi^*(G) \to G$ from the induced groupoid (Section 5.3) is a weak equivalence if and only if the submersion $\mathrm{t}\circ\mathrm{pr}_1$ is surjective.

Proposition 5.11 *Every strong equivalence of Lie groupoids is a weak equivalence.*

Proof Let $\phi\colon G \to H$ be a strong equivalence, with $\psi\colon H \to G$ and S and T as in the definition of strong equivalence above. We prove first that the map

$$\mathrm{t}\circ\mathrm{pr}_1\colon H_1 \times_{H_0} G_0 \longrightarrow H_0$$

of the definition of weak equivalence above is a surjective submersion. Clearly it is surjective because any $y \in H_0$ is the image of $(T(y), \psi(y))$. To see that it is a submersion, we prove that it has a local section through any point $(h_0\colon \phi(x_0) \to y_0, x_0)$ of $H_1 \times_{H_0} G_0$. To this end, consider the arrow

$$T(y_0)^{-1} h_0\colon \phi(x_0) \longrightarrow \phi(\psi(y_0))$$

in H. Since ϕ is an equivalence of categories, there is a unique arrow $g_0\colon x_0 \to \psi(y_0)$ in G with $\phi(g_0) = T(y_0)^{-1} h_0$. Let $\lambda\colon U \to G_1$ be a local bisection through g_0 in G, and let $\tilde{\lambda} = \mathrm{t}\circ\lambda\colon U \to G_0$ be the associated diffeomorphism onto an open neighbourhood V of $\psi(y_0)$. Let $\kappa\colon \psi^{-1}(V) \to H_1 \times_{H_0} G_0$ be the map

$$\kappa(y) = (T(y)\phi(\lambda(\tilde{\lambda}^{-1}(\psi(y)))), \tilde{\lambda}^{-1}(\psi(y))) \, .$$

Then κ is a section of $t \circ pr_1$ through the given point (h_0, x_0).

This proves that $t \circ pr_1$ is a surjective submersion. In particular, the fibred product $G_0 \times_{H_0} H_1 \times_{H_0} G_0$ of $t \circ pr_1$ along $\phi \colon G_0 \to H_0$ is a manifold, which fits into a pull-back diagram

$$
\begin{array}{ccc}
G_0 \times_{H_0} H_1 \times_{H_0} G_0 & \xrightarrow{\ pr_2\ } & H_1 \\
{\scriptstyle (pr_3, pr_1)} \downarrow & & \downarrow {\scriptstyle (s,t)} \\
G_0 \times G_0 & \xrightarrow{\ \phi \times \phi\ } & H_0 \times H_0
\end{array}
$$

Since ϕ is an equivalence of categories, the map $G_1 \to G_0 \times_{H_0} H_1 \times_{H_0} G_0$, sending g to $(s(g), \phi(g), t(g))$, is a bijection. We leave it to the reader to prove that it is in fact a diffeomorphism. $\qquad\Box$

We now list some basic properties of weak equivalences. We leave the elementary proofs to the reader.

Proposition 5.12 *Let G, H and K be Lie groupoids.*

(i) For two homomorphisms $\phi, \psi \colon G \to H$, if there is a transformation $T \colon \phi \to \psi$ then ϕ is a weak equivalence if and only if ψ is.

(ii) If for a weak equivalence $\psi \colon G \to H$ the map $t \circ pr_1$ of the condition (ES) has a section, then ϕ is a strong equivalence.

(iii) The composition of two weak equivalences is a weak equivalence.

(iv) For any weak equivalence $\phi \colon G \to H$ and any homomorphism $\psi \colon K \to H$, the weak pull-back

$$
\begin{array}{ccc}
P & \xrightarrow{\ pr_2\ } & K \\
{\scriptstyle pr_1} \downarrow & & \downarrow {\scriptstyle \psi} \\
G & \xrightarrow{\ \phi\ } & H
\end{array}
$$

exists and pr_2 is a weak equivalence for which $P_0 \to K_0$ is a surjective submersion.

REMARK. One says that two Lie groupoids G and G' are *weakly equivalent* (or *Morita equivalent*) if there exist weak equivalences $\phi \colon H \to G$ and $\phi' \colon H \to G'$ for a third Lie groupoid H. It follows from Proposition 5.12 that this defines an equivalence relation between Lie groupoids. Indeed, to check transitivity, suppose that we are given further weak equivalences $\psi \colon K \to G'$ and $\psi' \colon K \to G''$. Now form the weak pull-back P of ϕ' and ψ, and observe that G and G'' are weakly equivalent via $\phi \circ \mathrm{pr}_1$ and $\psi' \circ \mathrm{pr}_2$:

$$
\begin{array}{ccccc}
P & \xrightarrow{\ \mathrm{pr}_2\ } & K & \xrightarrow{\ \psi'\ } & G'' \\
{\scriptstyle\mathrm{pr}_1}\downarrow & & {\scriptstyle\psi}\downarrow & & \\
H & \xrightarrow{\ \phi'\ } & G' & & \\
{\scriptstyle\phi}\downarrow & & & & \\
G & & & &
\end{array}
$$

If G and G' are weakly equivalent Lie groupoids, we may in fact find a Lie groupoid H and weak equivalences $H \to G$ and $H \to G'$ which are surjective submersions on objects. To see this, assume that we have weak equivalences $H' \to G$ and $H' \to G'$. We now obtain H by the following diagram of weak pull-backs:

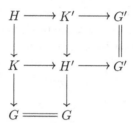

Many properties of Lie groupoids are stable under weak equivalence. One of them is Hausdorffness, as we will show in the next proposition. At this point we should remark that for this to be true, it is essential that we assumed in our definition of Lie groupoids that their base manifolds (of objects) are Hausdorff. In fact, one could consider more general Lie groupoids which may have non-Hausdorff base manifolds, but it is not difficult to see that any such Lie groupoid with non-Hausdorff base manifold is weakly equivalent to a Lie groupoid with Hausdorff base manifold: just take the induced groupoid on a Hausdorff open cover of the base manifold.

Proposition 5.13 *Let G and H be weakly equivalent Lie groupoids. Then G is Hausdorff if and only if H is.*

Proof By the preceding remark, we may assume that there is a weak equivalence $\phi\colon H \to G$ which is a surjective submersion on base manifolds, so we have the following pull-back diagram, with a surjective submersion $\phi \times \phi$ on the bottom:

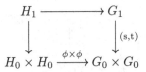

It is clear that if G_1 is Hausdorff then so is H_1. Conversely, assume that H_1 is Hausdorff. It is enough to show that for any $(x, y) \in G_0 \times G_0$ there exists an open neighbourhood $U \subset G_0 \times G_0$ such that $(s, t)^{-1}(U)$ is Hausdorff. But $\phi \times \phi \colon H_0 \times H_0 \to G_0 \times G_0$ is a surjective submersion, so we can choose U so small that $\phi \times \phi$ has a local section defined on U. Therefore $H_1 \to G_1$ has a local section defined on $(s, t)^{-1}(U)$ by the pull-back property of the diagram above, and because this section maps to a Hausdorff space, it follows that $(s, t)^{-1}(U)$ is Hausdorff. $\square$

The next property stable under weak equivalence is transitivity. Recall from Example 5.10 (3) that a Lie groupoid is called *transitive* if the map $(s, t) \colon G_1 \to G_0 \times G_0$ is a surjective submersion.

Proposition 5.14 *Let G be a Lie groupoid. The following conditions are equivalent.*

(i) G is transitive.

(ii) G is weakly equivalent to a Lie group.

(iii) $t \colon G(x_0, \text{-}) \to G_0$ is a surjective submersion for any (or some) $x_0 \in G_0$.

(iv) The inclusion $G_{x_0} \to G$ is a weak equivalence for any (or some) $x_0 \in G_0$.

(v) G is isomorphic to the gauge groupoid associated to a principal Lie group bundle (see Example 5.1 (8)).

Proof (i)$\Rightarrow$(iii) The map $t \colon G(x_0, \text{-}) \to G_0$ is the pull-back of the surjective submersion (s, t) along the map $G_0 \to G_0 \times G_0$ which sends x to (x_0, x), hence it is itself a surjective submersion.

(iii)$\Rightarrow$(iv) The group G_{x_0} is the restriction of G to $\{x_0\}$, so we only need to show that $t \circ \text{pr}_1 \colon G_1 \times_{G_0} \{x_0\} \to G_0$ is a surjective submersion. But this is exactly the map of (iii).

(iv)$\Rightarrow$(ii) This is trivial.

(ii)$\Rightarrow$(i) Since any Lie group is clearly transitive, we only need to show that transitivity is stable under weak equivalence. This is true because a pull-back of a surjective submersion is a surjective submersion, and because a map which pulls back along a surjective submersion to a surjective submersion is itself a surjective submersion.

(v)$\Rightarrow$(i) If $\pi\colon P \to M$ is a principal K-bundle, where K is a Lie group, then $(\mathrm{s},\mathrm{t})\colon \mathrm{Gauge}(P) \to M \times M$ is induced by the surjective submersion $(\pi,\pi)\colon P \times P \to M \times M$, and hence it is itself a surjective submersion. Therefore $\mathrm{Gauge}(P)$ is transitive.

(iii)$\Rightarrow$(v) By Theorem 5.4 we know that the map of (iii) is a principal G_{x_0}-bundle over G_0, so we only need to prove that the gauge groupoid of this bundle is isomorphic to G. For this, check that the map

$$G(x_0, \text{-}\,) \times G(x_0, \text{-}\,) \longrightarrow G\,,$$

which sends (g, g') to $g'g^{-1}$, induces an isomorphism

$$\mathrm{Gauge}(G(x_0, \text{-}\,)) \longrightarrow G$$

of Lie groupoids. $\square$

In an analogous way one can easily characterize the Lie groupoids which are weakly equivalent to discrete groups:

Proposition 5.15 *Let G be a Lie groupoid. The following conditions are equivalent.*

(i) $(\mathrm{s},\mathrm{t})\colon G_1 \to G_0 \times G_0$ *is a covering projection.*

(ii) G *is weakly equivalent to a discrete group.*

(iii) $\mathrm{t}\colon G(x_0, \text{-}\,) \to G_0$ *is a covering projection for any (or some) $x_0 \in G_0$.*

(iv) *The isotropy group G_{x_0} is discrete and the inclusion $G_{x_0} \to G$ is a weak equivalence, for any (or some) $x_0 \in G_0$.*

(v) *G is isomorphic to the gauge groupoid associated to a principal bundle with discrete structure group (i.e. a regular covering).*

Proof (i)$\Rightarrow$(iii) The map $\mathrm{t}\colon G(x_0, \text{-}\,) \to G_0$ is the pull-back of the covering projection (s,t) along $G_0 \to G_0 \times G_0$ which sends x to (x_0, x), so it is itself a covering projection.

(iii)$\Rightarrow$(iv) First, Theorem 5.4 implies that the isotropy group G_{x_0} is discrete. Now the result follows from Proposition 5.14.

(iv)$\Rightarrow$(ii) Trivial.

(ii)$\Rightarrow$(i) The condition (i) is stable under weak equivalence. This is because a pull-back of a covering projection is a covering projection, and because a map which pulls back along a surjective submersion to a covering projection is itself a covering projection.

The proofs of (v)$\Rightarrow$(i) and (iii)$\Rightarrow$(v) are also analogous to that of Proposition 5.14. $\qquad\qquad\square$

5.5 Etale groupoids

An *étale groupoid* is a Lie groupoid G with $\dim G_1 = \dim G_0$.

Recall that a smooth map $f\colon N \to M$ is a local diffeomorphism if $(df)_y$ is invertible for any $y \in N$ (thus in particular $\dim M = \dim N$). Equivalently, any $y \in N$ has an open neighbourhood $V \subset N$ such that $f|_V$ is an open embedding. A local diffeomorphism is also referred to as an *étale* map.

Exercises 5.16 (1) Show that a pull-back of a local diffeomorphism is a local diffeomorphism. Conversely, if the pull-back of a smooth map f along a surjective submersion is a local diffeomorphism, then f is itself a local diffeomorphism.

(2) Prove that a Lie groupoid G is étale if and only if the source map of G is a local diffeomorphism. In fact, all the structure maps of an étale groupoid are local diffeomorphisms.

(3) Show that for an étale groupoid the fibres of the source map, the fibres of the target map, the isotropy groups and the orbits are discrete.

(4) Prove that any weak equivalence between étale groupoids $G \to H$ is an étale map on objects and on arrows. (Hint: show that it is a submersion and that $\dim H = \dim G$.)

Examples 5.17 (1) The unit groupoid of a smooth manifold is étale.

(2) A discrete group is an étale groupoid over a one point space.

(3) If G is a discrete group acting on a manifold M, the associated action groupoid $G \ltimes M$ is étale.

(4) Let G be a Lie groupoid. The Lie groupoid of local bisections $\mathrm{Bis}(G)$ (Section 5.1) is an étale groupoid. It is isomorphic to G if and only if G is itself étale.

(5) Let M be a manifold. The germs of locally defined diffeomorphisms $f\colon U \to V$ between open subsets of M form a groupoid $\Gamma(M)$ over M; the germ of f at x is an arrow from x to $f(x)$, and the multiplication in $\Gamma(M)$ is induced by the composition of diffeomorphisms. There are a natural sheaf topology and a smooth structure on $\Gamma(M)_1$ such that the structure maps of $\Gamma(M)$ are étale. Thus $\Gamma(M)$ is an étale groupoid. In particular, the étale groupoid $\Gamma(\mathbb{R}^q)$ is referred to as the *Haefliger groupoid*, and denoted by Γ^q.

Exercise 5.18 Show that for any manifold M of dimension q, the étale groupoid $\Gamma(M)$ is weakly equivalent to the Haefliger groupoid Γ^q.

Example 5.19 (Etale holonomy and monodromy groupoids) Let $\mathcal{F}$ be a foliation of a manifold M. We can choose a *complete transversal section* S of $(M, \mathcal{F})$, i.e. an immersed (not necessarily connected) submanifold of M of dimension equal to the codimension of $\mathcal{F}$, which is transversal to the leaves of $\mathcal{F}$ and intersects any leaf in at least one point. For instance, one can take S to be the union of a countable disjoint family of (local) transversal sections. Denote by $\iota\colon S \to M$ the inclusion.

We can now define a Lie groupoid $\mathrm{Mon}_S(M, \mathcal{F})$ over S as the induced groupoid

$$\mathrm{Mon}_S(M, \mathcal{F}) = \iota^*(\mathrm{Mon}(M, \mathcal{F})) \,.$$

This groupoid can be seen as the restriction of the (full) monodromy groupoid $\mathrm{Mon}(M, \mathcal{F})$ to S: the arrows of $\mathrm{Mon}_S(M, \mathcal{F})$ are those arrows of $\mathrm{Mon}(M, \mathcal{F})$ which start and end in the submanifold S. The induced groupoid $\iota^*(\mathrm{Mon}(M, \mathcal{F}))$ is indeed a Lie groupoid because the composition $\mathrm{t} \circ \mathrm{pr}_1\colon \mathrm{Mon}(M, \mathcal{F})_1 \times_M S \to M$ is a surjective local diffeomorphism (Section 5.3).

In particular, note that $\dim \mathrm{Mon}_S(M, \mathcal{F})_1 = \dim S$, so the Lie groupoid $\mathrm{Mon}_S(M, \mathcal{F})$ is étale. It is referred to as the *étale monodromy groupoid* over S associated to $(M, \mathcal{F})$. The inclusion of $\mathrm{Mon}_S(M, \mathcal{F})$ into $\mathrm{Mon}(M, \mathcal{F})$ is a weak equivalence. For any point $x \in S$ on a leaf L of $\mathcal{F}$ we have

$$\mathrm{Mon}_S(M, \mathcal{F})_x = \mathrm{Mon}(M, \mathcal{F})_x = \pi_1(L, x) \,.$$

In a completely analogous way we now define the *étale holonomy groupoid* $\mathrm{Hol}_S(M, \mathcal{F})$ over S, weakly equivalent to $\mathrm{Hol}(M, \mathcal{F})$, by

$$\mathrm{Hol}_S(M, \mathcal{F}) = \iota^*(\mathrm{Hol}(M, \mathcal{F})) \,.$$

Note that for any $x \in S$ we have

$$\mathrm{Hol}_S(M, \mathcal{F})_x = \mathrm{Hol}(M, \mathcal{F})_x = \mathrm{Hol}(L, x) \,,$$

where L is the leaf of $\mathcal{F}$ through x.

The same idea of restricting a Lie groupoid to a complete transversal leads to the following characterization of étale groupoids up to weak equivalence.

Proposition 5.20 *A Lie groupoid is weakly equivalent to an étale one if and only if it has discrete isotropy groups.*

Proof Any étale groupoid has discrete isotropy groups, and this property is preserved under weak equivalence. Conversely, assume that G is a Lie groupoid of dimension m with discrete isotropy groups. Let n be the dimension of G_0. Theorem 5.4 shows that all the orbits of G are of the same dimension $m - n$; furthermore, they form a foliation of G_0. Indeed, the vectors tangent to the orbits in G_0 are exactly the images along dt of vectors in $\mathrm{Ker}(ds) \subset T(G_1)$. Since any section of $\mathrm{Ker}(ds)$ defined on G_0 can be extended to all of G_1 by right translations, and the obtained vector field is projectable along dt, this implies that $(dt)(\mathrm{Ker}(ds))$ is an involutive subbundle of $T(G_0)$.

Now take a complete transversal S of the foliation of G_0 by orbits of G. First we will prove that the map

$$t\colon s^{-1}(S) \cong G_1 \times_{G_0} S \longrightarrow G_0$$

is a surjective local diffeomorphism. To see this, first note that the dimension of $s^{-1}(S)$ is n, so it is enough to show that this map is a surjective immersion. Let $g\colon x \to y$ be an arrow on G. Note first that $G(x, y)$ is also discrete, because it is diffeomorphic to the isotropy group G_x via translation by g. Now take any $\xi \in T_g(s^{-1}(S))$ with $dt(\xi) = 0$. It follows that $ds(\xi)$ is tangent to the foliation of G_0 by orbits, and since it is also tangent to S, we must have $ds(\xi) = 0$. Thus ξ is tangent to the discrete space $G(x, y)$, hence $\xi = 0$. Finally, the map $t\colon s^{-1}(S) \to G_0$ is surjective since S is a complete transversal.

With this, it follows that the restriction $G|_S$ of G to S is a Lie groupoid weakly equivalent to G. It also follows that this groupoid is étale, because it can be obtained as the inverse image of S along the étale map $t\colon s^{-1}(S) \to G_0$.   □

Let G be an étale groupoid. There is a canonical homomorphism of Lie groupoids

$$\mathrm{Eff}\colon G \longrightarrow \Gamma(G_0) \, ,$$

which is the identity on objects and is given on arrows by

$$\mathrm{Eff}(g) = \mathrm{germ}_{s(g)}(t \circ (s|_U)^{-1}) \, ,$$

where $g \in G_1$ and U is any open neighbourhood of g in G_1 such that both $s|_U$ and $t|_U$ are injective. The map $\mathrm{Eff}\colon G_1 \to \Gamma(G_0)_1$ is a local diffeomorphism. An *effective groupoid* is an étale groupoid for which the homomorphism Eff is injective (on arrows). The image $\mathrm{Eff}(G)$ of Eff is an open subgroupoid of $\Gamma(G_0)$ and hence effective; it is referred to as the *effect* of G.

Examples 5.21 (1) Let $\mathcal{F}$ be a foliation of M, and let S be a complete transversal of $\mathcal{F}$. Recall that for any two points $x, y \in S$ on the same leaf L of $\mathcal{F}$, the arrows in $\mathrm{Mon}_S(M, \mathcal{F})$ from x to y are the homotopy classes of paths from x to y inside L. But the holonomy class of such a path α may be faithfully represented by the germ of a locally defined diffeomorphism on S, namely by $\mathrm{hol}^{S,S}(\alpha)$. It follows that the effect homomorphism of the étale holonomy groupoid $\mathrm{Mon}_S(M, \mathcal{F})$ is given by the holonomy $\mathrm{Eff} = \mathrm{hol}^{S,S} \colon \mathrm{Mon}_S(M, \mathcal{F}) \to \Gamma(S)$, and

$$\mathrm{Eff}(\mathrm{Mon}_S(M, \mathcal{F})) = \mathrm{Hol}_S(M, \mathcal{F}) \, .$$

In particular, the groupoid $\mathrm{Hol}_S(M, \mathcal{F})$ is effective.

(2) The class of effective groupoids is stable under weak equivalence among étale groupoids. In other words, if two étale groupoids are weakly equivalent, then one is effective if and only if the other is too.

(3) Let $f \colon M \to N$ be a surjective submersion with connected fibres, and $\mathcal{F}$ the associated foliation of M. Then the étale holonomy groupoid of $(M, \mathcal{F})$ is weakly equivalent to the manifold N regarded as a unit groupoid.

(4) Let G be a discrete group acting freely and properly discontinuously (from the right) on a connected manifold $\tilde{M}$ such that $\tilde{M}/G = M$ is a Hausdorff manifold. Suppose that we also have a left action of G on a manifold F. Consider the associated foliation $\mathcal{F}$ on the flat bundle $\tilde{M} \times_G F$ over M. Take $x_0 \in \tilde{M}$. Let S be the complete transversal section of $\mathcal{F}$ given by the image of $\{(x_0, z) \mid z \in F\}$ under the quotient map $\tilde{M} \times F \to \tilde{M} \times_G F$. Then we have

$$\mathrm{Hol}_S(\tilde{M} \times_G F, \mathcal{F}) \cong \mathrm{Eff}(G \ltimes F) \, .$$

(5) Let $\mathcal{F}$ be the standard foliation of the Möbius band M. By (4) just above, the étale holonomy groupoid of $(M, \mathcal{F})$ is isomorphic to the translation groupoid $\mathbb{Z}_2 \ltimes (-1, 1)$.

(6) The Kronecker foliation $\mathcal{F}$ of torus T^2 is also a suspension, so the étale holonomy groupoid is $\mathbb{Z} \ltimes S^1$.

(7) Let $\mathcal{F}$ be the Reeb foliation of S^3. We may choose a transversal section S of $\mathcal{F}$ so that the associated étale holonomy groupoid is isomorphic to $(\mathbb{Z} \oplus \mathbb{Z}) \ltimes \mathbb{R}$.

Exercises 5.22 (1) Let

be a weak pull-back of Lie groupoids. Show that if K and H are étale, then P is also étale.

(2) Let G and G' be weakly equivalent Lie groupoids. We know that there exist a Lie groupoid H and weak equivalences $\phi\colon H \to G$ and $\phi'\colon H \to G'$ which are surjective submersions on objects; show that H can be chosen étale if G and G' are.

Examples 5.23 (1) Let M be a smooth manifold. A *(local) transition* on M is a diffeomorphism $f\colon U \to U'$ between two open subsets of M. We shall denote the set of all transitions on M by C^∞_M. A *pseudogroup* on M is a subset P of transitions on M such that

  (i) $\mathrm{id}_U \in P$ for any open $U \subset M$,
 (ii) if $f, f' \in P$, then $f' \circ f|_{f^{-1}(\mathrm{dom}(f'))} \in P$ and $f^{-1} \in P$, and
(iii) if f is a transition on M and (U_i) is an open cover of $\mathrm{dom}(f)$ such that $f|_{U_i} \in P$ for any i, then $f \in P$.

In particular, the collection C^∞_M of all transitions on M is a pseudogroup on M.

To any pseudogroup P on M we can associate an effective groupoid $\Gamma(P)$ over M as follows: for any $x, y \in M$ let

$$\Gamma(P)(x,y) = \{\mathrm{germ}_x f \mid f \in P,\ x \in \mathrm{dom}(f),\ f(x) = y\}\,.$$

The multiplication in $\Gamma(P)$ is given by the composition of transitions. With the classical sheaf topology $\Gamma(P)_1$ becomes a smooth manifold (which may be neither Hausdorff nor second-countable), and $\Gamma(P)$ becomes an effective groupoid. The effective groupoid $\Gamma(M)$ (Example 5.17 (5)) is a special case of this construction, since $\Gamma(M) = \Gamma(C^\infty_M)$.

There is also a natural way to construct a pseudogroup out of an étale groupoid. For any étale groupoid G, put

$$\Psi(G) = \{t \circ \sigma \mid \sigma \text{ is a local bisection of } G\}\,.$$

This is clearly a pseudogroup on G_0. We have

$$\Gamma(\Psi(G)) = \mathrm{Eff}(G)\,, \qquad \Psi(\Gamma(P)) = P$$

for any étale groupoid G and any pseudogroup P. In particular, we may identify effective groupoids over M with pseudogroups on M.

(2) A pseudogroup P on M is called finitely (countably) generated if there exists a finite (countable) subset $A \subset P$ such that P is the smallest pseudogroup on M which contains A (i.e. P is generated by A). An example of a countably generated pseudogroup is the *holonomy pseudogroup* of a foliated manifold $(M, \mathcal{F})$ on a complete transversal section S, i.e. the pseudogroup associated to the étale holonomy groupoid of $(M, \mathcal{F})$. In fact, this pseudogroup is finitely generated if M is compact.

Conversely, any countably generated pseudogroup is the holonomy pseudogroup of a foliation. Indeed, let P be a pseudogroup on a manifold N of dimension q, generated by a countable subset $(f_i)_{i=1}^{\infty}$. Let V be the open subset of $N \times \mathbb{R} \times \mathbb{R}$ given by

$$V = (N \times \mathbb{R} \times (0,1)) \cup \bigcup_{i=1}^{\infty} (\mathrm{dom}(f_i) \times (i, i+1) \times (0,3)) .$$

There is a natural foliation on V of codimension q given by the first projection. Let M be the manifold of dimension $q + 2$, obtained as the quotient of V by identifying (y, t, t') with $(f_i(y), t, t' - 2)$ for any $i = 1, 2, \ldots$, $y \in \mathrm{dom}(f_i)$, $t \in (i, i+1)$ and $t' \in (2, 3)$. Observe that the foliation on V induces a foliation $\mathcal{F}$ on M of codimension q. For any $y \in N$ denote by $T(y) \in M$ the equivalence class of the point $(y, 0, 1/2) \in V$. The image of $T \colon N \to M$, which is isomorphic to N, is a complete transversal section of $\mathcal{F}$. It is easy to check that the pseudogroup of $(M, \mathcal{F})$ on this complete transversal section is isomorphic to P. (The idea for this construction was communicated to us by J. Pradines, who attributed it to G. Hector.)

It is an open problem to characterize finitely generated pseudogroups which come from the foliations of compact manifolds.

Exercise 5.24 Let P be a pseudogroup on M and P' a pseudogroup on M'. A *(local) transition* from M to M' is a diffeomorphism $h \colon V \to V'$ from an open subset V of M to an open subset V' of M'. An *equivalence* from P to P' is a subset R of transitions from M to M' such that

(i) $\bigcup_{h \in R} \mathrm{dom}(h) = M$ and $\bigcup_{h \in R} \mathrm{cod}(h) = M'$,

(ii) for any $h, k \in R$, $f \in P$ and $f' \in P'$ it holds that $h \circ f \circ k^{-1} \in P'$, $h^{-1} \circ f' \circ k \in P$ and $f' \circ h \circ f \in R$, and

(iii) R is a maximal family of transitions from M to M' satisfying (i) and (ii).

Two pseudogroups P on M and P' on M' are *equivalent* if there exists an equivalence between them. Show that pseudogroups P and P' are equivalent if and only if $\Gamma(P)$ and $\Gamma(P')$ are weakly equivalent.

5.6 Proper groupoids and orbifolds

In this section we shall consider another class of Lie groupoids which is stable under weak equivalence: proper groupoids.

First, recall that a map between (Hausdorff) manifolds $f\colon N \to M$ is *proper* if $f^{-1}(K)$ is compact for any compact subset K of M. In particular, any proper map between (Hausdorff) manifolds is closed. Before we present the definition of proper groupoids, we begin with a basic lemma concerning the behaviour of proper maps between manifolds under pull-backs. We omit the proof, which is elementary.

Lemma 5.25 *Let $g\colon N' \to M'$ be the pull-back of a map $f\colon N \to M$ along a smooth map $h\colon M' \to M$, where M and M' are Hausdorff:*

$$
\begin{array}{ccc}
N' & \longrightarrow & N \\
{\scriptstyle g}\downarrow & & \downarrow{\scriptstyle f} \\
M' & \stackrel{h}{\longrightarrow} & M
\end{array}
$$

If N is Hausdorff and f is proper, then N' is also Hausdorff and g is proper. If, conversely, N' is Hausdorff, g proper and h a surjective submersion, then N is Hausdorff and f is proper.

A Lie groupoid G is said to be *proper* if it is Hausdorff and the map $(s,t)\colon G_1 \to G_0 \times G_0$ is proper. The stability of propriety under weak equivalence is an immediate consequence of the preceding lemma.

Proposition 5.26 *If G and H are weakly equivalent Lie groupoids, and if one of them is proper, then so is the other.*

Proposition 5.27 *Let*

$$
\begin{array}{ccc}
P & \longrightarrow & H \\
\downarrow & & \downarrow \\
K & \longrightarrow & G
\end{array}
$$

be a weak pull-back of Lie groupoids. If K and H are proper and G is Hausdorff, then P is also proper.

Proof The map $P_1 \to P_0 \times P_0$ is of the form

$$H_1 \times_{G_0} G_1 \times_{G_0} K_1 \longrightarrow (H_0 \times_{G_0} G_1 \times_{G_0} K_0) \times (H_0 \times_{G_0} G_1 \times_{G_0} K_0) .$$

After reshuffling the factors, we can write this map as a composition of pull-backs of $H_1 \to H_0 \times H_0$, $K_1 \to K_0 \times K_0$ and the diagonal $G_1 \to G_1 \times G_1$. All these maps are proper by assumption. Since proper maps are obviously closed under composition, the proposition is proved.

$\square$

Examples 5.28 (1) Let G be a Lie group acting on a manifold N. Then the translation groupoid $G \ltimes N$ is proper if and only if the action of G is proper in the classical sense (by definition). For example, if G is compact then $G \ltimes N$ is always proper.

(2) Let $\mathcal{F}$ be a foliation of a manifold M, and assume that any leaf of $\mathcal{F}$ is compact with finite holonomy (for instance, this is true if $\mathcal{F}$ is given by a foliated action of a compact Lie group, see Proposition 2.8). Then the holonomy groupoid $\mathrm{Hol}(M, \mathcal{F})$ is proper. In fact, more is true: the source map of the holonomy groupoid is proper, and the same is therefore true for the target map. To see this, we first use the local Reeb stability theorem to conclude that locally the foliation has the structure of a flat bundle (suspension), and then use Example 5.8 (3).

We shall now show that proper effective groupoids may be identified with orbifolds.

Let Q be an orbifold and $\mathcal{U} = \{(U_i, G_i, \phi_i)\}_{i \in I}$ an orbifold atlas of Q. Put $U = \coprod_{i \in I} U_i$ and $\phi = \{\phi_i\} \colon U \to Q$. Now let $\Psi(\mathcal{U})$ be the pseudogroup on U of all transitions f on U for which $\phi \circ f = \phi|_{\mathrm{dom}(f)}$. Define the effective groupoid $\Gamma(\mathcal{U})$ associated to the orbifold atlas $\mathcal{U}$ to be the effective groupoid associated to the pseudogroup $\Psi(\mathcal{U})$,

$$\Gamma(\mathcal{U}) = \Gamma(\Psi(\mathcal{U})) .$$

Proposition 5.29 *Let $\mathcal{U}$ be an orbifold atlas of an orbifold Q.*

(i) $\Gamma(\mathcal{U})$ is a proper effective groupoid.

(ii) If $\mathcal{U}$ is an orbifold atlas of an orbifold Q and $\mathcal{U}'$ is an orbifold atlas of an orbifold Q', then $\Gamma(\mathcal{U})$ and $\Gamma(\mathcal{U}')$ are weakly equivalent if and only if Q and Q' are isomorphic.

Proof (i) Take any $(x, y) \in U_i \times U_j \subset U \times U$. It is enough to find a compact neighbourhood K of (x, y) such that $(\mathrm{s}, \mathrm{t})^{-1}(K)$ is compact.

If $\phi_i(x) \neq \phi_j(y)$ this is easy since Q is locally compact and Hausdorff. So assume that $\phi_i(x) = \phi_j(y) = q$. As the charts of the atlas are compatible, it follows by Proposition 2.13 that there is an embedding $f\colon Z \to U_j$ of an open neighbourhood $Z \subset U_i$ of x with $f(x) = y$ and $\phi_j \circ f = \phi_i|_Z$. We may assume that Z is G_i-stable and $(G_i)_Z = (G_i)_x$. This implies that $f(Z)$ is G_j-stable and $(G_j)_{f(Z)} = (G_j)_y$. By Lemma 2.11 we see that

$$(\mathrm{s},\mathrm{t})^{-1}(Z \times f(Z)) = \{\mathrm{germ}_z(g \circ f) \,|\, g \in (G_j)_y,\, z \in Z\} \cong (G_j)_y \times Z \,.$$

The rest follows from the fact that $(G_j)_y$ is finite.

(ii) We mentioned that $\Gamma(\mathcal{U})$ and $\Gamma(\mathcal{U}')$ are weakly equivalent if and only if $\Psi(\mathcal{U})$ and $\Psi(\mathcal{U}')$ are equivalent, i.e. if there exists an equivalence between them. Such an equivalence clearly induces an isomorphism between the orbifolds, while conversely, for any isomorphism between orbifolds the local lifts (which must exists by definition, see Section 2.4) form an equivalence between the pseudogroups (see Exercise 5.24). $\qquad\Box$

This proposition implies, in particular, that for an orbifold Q the weak equivalence class of $\Gamma(\mathcal{U})$ is independent of the choice of the orbifold atlas $\mathcal{U}$ of Q. So we can associate to Q a proper effective groupoid

$$\Gamma(Q) \,,$$

which is determined uniquely up to weak equivalence.

In fact, any proper effective groupoid G comes from an orbifold in this way. To see this, note first that any isotropy group of a proper étale groupoid is finite. Furthermore, a proper étale groupoid locally looks like the translation groupoid with respect to an action of an isotropy group, as expressed in the following proposition.

Proposition 5.30 *Let G be a proper étale groupoid. Then any $x \in G_0$ has an open neighbourhood U in G_0 with an action of the isotropy group G_x such that there is an isomorphism of étale groupoids*

$$G|_U \cong G_x \ltimes U \,.$$

Proof Since $\theta = (\mathrm{s},\mathrm{t})\colon G_1 \to G_0 \times G_0$ is proper and G is étale, the groups $G_x = \theta^{-1}(x,x)$ are finite. By the argument given just before Theorem 2.9, we can find a connected open neighbourhood $W \subset G_0$ of x and sections $\sigma_g\colon W \to G_1$ (for $g \in G_x$) of the source map with $\sigma_g(x) = g$ such that $f_g = \mathrm{t} \circ \sigma_g$ is a diffeomorphism from W onto W and $f_g \circ f_h = f_{gh}$ for any $g, h \in G_x$. Thus $H = \{f_g \,|\, g \in G_x\}$ is a subgroup of $\mathrm{Diff}(W, x)$, and the map $g \to f_g$ gives us an action of G_x on W. Furthermore observe that $\sigma_g(W) \cap \sigma_{g'}(W) = \emptyset$ for any $g \neq g'$ because G is Hausdorff and connected. Since θ is proper and hence closed, $\theta(G_1 - \bigcup_{g \in G_x} \sigma_g(W))$ is closed in $G_0 \times G_0$. Thus we can find an open neighbourhood $U \subset W$ of x such that $U \times U$ is disjoint from $\theta(G_1 - \bigcup_{g \in G_x} \sigma_g(W))$. We can also choose U so small that U is H-stable. It is then clear that $G|_U$ is isomorphic to the translation groupoid $G_x \ltimes U$. $\square$

This proposition implies that any proper étale groupoid defines an orbifold structure on its space of orbits. The *space of orbits* G_0/G of a Lie groupoid G is defined as the space of orbits of the canonical right G-action on G_0. Equivalently, G_0/G is the quotient space of G_0 in which two points of G_0 are identified precisely if there exists an arrow in G between them.

Corollary 5.31 *Let G be a proper étale groupoid. Then there is a canonical orbifold structure on G_0/G such that $\Gamma(G_0/G)$ is weakly equivalent to $\mathrm{Eff}(G)$.*

Proof For any $x \in G_0$ we take an open neighbourhood U_x of x as in Proposition 5.30. We may also take U_x so small that it is diffeomorphic to an open subset V_x of $\mathbb{R}^n$, with diffeomorphism $\phi_x \colon U_x \to V_x$. Let H_x be the image of the action of G_x in $\mathrm{Diff}(U_x, x)$. Now

$$(U_x, \phi_x \circ H \circ \phi_x^{-1}, \pi \circ \phi_x^{-1})$$

is an orbifold chart on G_0/G with $\pi(x) \in \pi(V)$, where $\pi \colon G_0 \to G_0/G$ is the quotient projection. These charts form an orbifold atlas of G_0/G. $\square$

Theorem 5.32 *The following conditions are equivalent for any Lie groupoid G.*

(i) *G is weakly equivalent to a proper effective groupoid.*

(ii) *G is weakly equivalent to the holonomy groupoid of a foliation with compact leaves and finite holonomy groups.*

(iii) *G is weakly equivalent to the translation groupoid of a compact Lie group action, for which any isotropy group is finite and acts effectively on a slice.*

(iv) *G is weakly equivalent to the effective groupoid associated to an orbifold.*

Proof First, the equivalence (i)⇔(iv) follows from our discussion above (Proposition 5.29 and Corollary 5.31). By Proposition 2.23, any orbifold Q is isomorphic to the orbifold associated to the foliated action of the compact Lie group $U(n)$ on the unitary frame bundle $UF(Q)$, after we choose a Riemannian metric on Q. This action has finite isotropy groups and is effective on slices, as in (iii). In particular, this action is foliated and the associated foliation has compact leaves with finite holonomy groups, as in (ii). Furthermore, we saw that the space of leaves of the foliation and the space of orbits of the action are both orbifolds isomorphic to Q. So we only need to show that the associated Lie groupoids – the holonomy groupoid, the translation groupoid and the groupoid associated to Q – are weakly equivalent. To see this, choose for any point $x \in UF(Q)$ a slice, i.e. a transversal section S_x of the foliation, on which there is an action by the isotropy $U(n)_x$ such that the saturation of S_x is isomorphic to $S_x \times_{U(n)_x} U(n)$ (this is in fact the local Reeb stability theorem for the associated foliation). These local slices form an orbifold atlas of Q. Furthermore, by restricting the translation groupoid to the union of local slices we obtain a weakly equivalent groupoid which is exactly the proper effective groupoid associated to the orbifold atlas of local slices. This proves (iv)⇒(iii), while a similar argument shows also (iii)⇒(ii) and (ii)⇒(iv). □

Example 5.33 An orbifold is a manifold if and only if all the isotropy groups are trivial. Therefore a Lie groupoid G is weakly equivalent to (the unit groupoid of) a manifold if and only if it is a proper groupoid with trivial isotropy groups (such a groupoid is weakly equivalent to a proper étale one by Proposition 5.20).

5.7 Principal bundles over Lie groupoids

To conclude this chapter, we introduce actions of Lie groupoids, and associated connections. These notions will be used in Section 6.3 below.

Let G be a Lie groupoid. A G-bundle over a manifold M is a manifold P equipped with a map $\pi\colon P \to M$ and a smooth right G-action μ on P along $\epsilon\colon P \to G_0$ (see Section 5.3) which is fibrewise with respect to

π, i.e. $\pi(pg) = \pi(g)$ for any $p \in P$ and any $g \in G_1$ with $\epsilon(p) = \mathrm{t}(g)$.

$$P \xrightarrow{\ \epsilon\ } G_0$$
$$\pi \downarrow$$
$$M$$

Such a bundle P is said to be *principal* if

(i) π is a surjective submersion, and
(ii) the map $(\mathrm{pr}_1, \mu)\colon P \times_{G_0} G_1 \to P \times_M P$, sending (p, g) to (p, pg), is a diffeomorphism.

Note that in case G is a Lie group we recover the usual notion of a principal G-bundle.

For a principal G-bundle $\pi\colon P \to M$, we refer to the manifold P as the *total space* of the bundle, and we shall denote by $\delta\colon P \times_M P \to G_1$ the map $\mathrm{pr}_2 \circ (\mathrm{pr}_1, \mu)^{-1}$. This map is uniquely determined by the identity $p\delta(p, p') = p'$ and satisfies the equation $\delta(p, p')g = \delta(p, p'g)$.

An *equivariant map* between principal G-bundles $\pi\colon P \to M$ and $\pi'\colon P' \to M$ over M is a smooth map $f\colon P \to P'$ which commutes with all the structure maps, i.e. the identities $\pi'(f(p)) = \pi(p)$, $\epsilon'(f(p)) = \epsilon(p)$ and $f(pg) = f(p)g$ hold, for any $p \in P$ and $g \in G_1$ with $\epsilon(p) = \mathrm{t}(g)$.

Remarks 5.34 (1) The space G_1 of arrows of a Lie groupoid G carries the structure of a principal G-bundle over G_0: for π one takes the target map and for ϵ the source map, while the right action is given by the multiplication in G. We call this bundle the *unit* bundle of G, and denote it by $U(G)$.

(2) If P is a principal G-bundle over M and $f\colon N \to M$ is a smooth map, the pull-back $N \times_M P$ has the structure of a principal G-bundle over N. We denote this bundle by $f^*(P)$.

(3) Combining the previous two remarks, we see that for any map $\alpha\colon M \to G_0$ there is a principal G-bundle $\alpha^*(U(G))$ over M. Its total space is the space of pairs (m, g) where g is an arrow with target $\alpha(m)$. Bundles of this form are called *trivial*.

(4) Let P be a principal G-bundle over M. Take any point $m \in M$, and choose a local section $\sigma\colon U \to P$ of π defined on an open neighbourhood U of m. Let $\alpha = \epsilon \circ \sigma\colon U \to G_0$. Then the map $\alpha^*(U(G)) \to P$, which sends (m, g) to $\sigma(m)g$, is an isomorphism from the trivial bundle $\alpha^*(U(G))$ to the restriction $P_U = \pi^{-1}(U)$. Its inverse sends p to $\delta(\sigma(\pi(p)), p)$. Thus, any principal bundle is locally trivial.

(5) Every equivariant map $P \to P'$ between principal G-bundles over M is an isomorphism. In fact by (4) it is sufficient to check this for trivial bundles. But for any $\alpha, \beta\colon M \to G_0$, a map between trivial bundles $f\colon \alpha^*(U(G)) \to \beta^*(U(G))$ is completely determined by the map $\phi\colon M \to G_1$ sending m to $\mathrm{pr}_2(f(m, 1_{\alpha(m)}))$, since

$$f(m, g) = f(m, 1_{\alpha(m)})g = (m, \phi(m)g) .$$

Thus clearly f is an isomorphism, with the inverse given by $f^{-1}(m, g) = (m, \phi(m)^{-1}g)$.

Lemma 5.35 *Let $\pi\colon P \to M$ be a principal G-bundle, let Q be a manifold with a right G-action, and let $f\colon Q \to P$ be a submersion preserving the G-action. Then Q/G is a manifold and the quotient map $Q \to Q/G$ is a principal G-bundle.*

Proof Let $q \in Q$, and choose a neighbourhood U of $m = \pi(f(q))$ so small that $U \subset \pi(f(Q))$ and that π has a section $\sigma\colon U \to P$ with $\sigma(m) = f(q)$. Then the restriction P_U is trivial and f maps $Q_U = f^{-1}(P_U)$ onto P_U. Since the conclusion of the lemma is local in M, this shows that we may assume that P is the trivial bundle given by a map $\alpha\colon M \to G_0$ and that f is a surjective submersion,

$$f\colon Q \longrightarrow P = \alpha^*(U(G)) = M \times_{G_0} G_1 .$$

Denote by uni: $G_0 \to G_1$ the unit map of G. Let $R \to M$ be the pull-back of f along the embedding $(\mathrm{id}, \mathrm{uni} \circ \alpha)\colon M \to M \times_{G_0} G_1$, mapping m into $(m, 1_{\alpha(m)})$. Thus R may be identified with the submanifold $f^{-1}(M \times_{G_0} G_0)$ of Q. Consider the map $\rho\colon Q \to R$ defined by

$$\rho(q) = q(\mathrm{pr}_2(f(q)))^{-1} .$$

This map is a submersion because it is also isomorphic to the pull-back of the target map of G along the map $\alpha \circ \mathrm{pr}_1\colon R \to G_0$,

$$
\begin{array}{ccccc}
Q & \xrightarrow{\ f\ } & P & \xrightarrow{\ \mathrm{pr}_2\ } & G_1 \\
\Updownarrow{\scriptstyle \rho} & & \Updownarrow{\scriptstyle \pi} & & \downarrow{\scriptstyle t} \\
R & \xrightarrow[\ \mathrm{pr}_1\ ]{} & M & \xrightarrow[\ \alpha\ ]{} & G_0
\end{array}
$$

and Q becomes a (trivial) principal G-bundle over R. Indeed, for any two points $q_1, q_2 \in Q$ with $\rho(q_1) = \rho(q_2)$ we have

$$q_2 = q_1(\mathrm{pr}_2(f(q_1)))^{-1}(\mathrm{pr}_2(f(q_2))) ,$$

so the map $Q \times_{G_0} G_1 \to Q \times_R Q$ has the inverse which maps (q_1, q_2) to $(q_1, \mathrm{pr}_2(f(q_1))^{-1}\mathrm{pr}_2(f(q_2)))$. $\qquad\square$

Let G and H be Lie groupoids. A principal G-bundle over H is a principal G-bundle $\pi \colon P \to H_0$ over the manifold H_0,

$$
\begin{array}{ccc}
P & \xrightarrow{\ \epsilon\ } & G_0 \\
{\scriptstyle \pi}\big\downarrow & & \\
H_0 & &
\end{array}
$$

which is equipped with a left H-action on P along π, which commutes with the right G-action, i.e. $\epsilon(hp) = \epsilon(p)$ and

$$(hp)g = h(pg)$$

for any $h \in H_1$, $p \in P$ and $g \in G_1$ with $\mathrm{s}(h) = \pi(p)$ and $\epsilon(p) = \mathrm{t}(g)$.

A map $P \to P'$ between principal G-bundles over H is a map of principal G-bundles over H_0 which also respects the H-action. As we have seen, any such map is an isomorphism.

Example 5.36 Let G be a Lie group. Then for any foliated manifold $(M, \mathcal{F})$ the principal G-bundles over $\mathrm{Mon}(M, \mathcal{F})$ are the foliated principal G-bundles over M (Subsection 4.2.2), while the principal G-bundles over $\mathrm{Hol}(M, \mathcal{F})$ are the transverse principal G-bundles over M. If H is another Lie group, the principal G-bundles over H are conjugacy classes of homomorphisms $H \to G$.

Exercise 5.37 Let H be a Lie groupoid. A vector bundle E over H is defined to be a (real) vector bundle $\pi \colon E \to H_0$ over the space H_0 of objects, equipped with a left action of H along π, which is linear as map between the fibres of π. A metric on such a vector bundle is a metric in the usual sense, which is preserved by the action of H.

(1) Show that if H is an étale groupoid, then the tangent bundle of H_0 has the natural structure of a vector bundle over H.

Consider now the case where H is the holonomy groupoid $\mathrm{Hol}(M, \mathcal{F})$ of a foliated manifold $(M, \mathcal{F})$. Let T be a complete transversal of $(M, \mathcal{F})$, with the associated étale holonomy groupoid $\mathrm{Hol}_T(M, \mathcal{F})$.

(2) Show that the normal bundle $N(\mathcal{F})$ of the foliation $\mathcal{F}$ has the natural structure of a vector bundle over $\mathrm{Hol}(M, \mathcal{F})$. Show that this bundle carries a metric if and only if the foliation is Riemannian. Show

that the bundle is trivial (as a bundle with H-action) if and only if the foliation is transversely parallelizable.

(3) Show that the pull-back of the normal bundle $N(\mathcal{F})$ along the inclusion $\mathrm{Hol}_T(M, \mathcal{F}) \to \mathrm{Hol}(M, \mathcal{F})$ is isomorphic to the tangent bundle of $\mathrm{Hol}_T(M, \mathcal{F})$. Prove that $(M, \mathcal{F})$ is Riemannian if and only if the tangent bundle of $\mathrm{Hol}_T(M, \mathcal{F})$ carries a metric, and that $(M, \mathcal{F})$ is transversely parallelizable if and only if this tangent bundle is trivial.

Let G be a Lie groupoid, and let $\pi\colon P \to B$ be a principal G-bundle over a manifold B along $\epsilon\colon P \to G_0$. For any $p \in P$, denote by $\mathcal{V}_p$ the space $\mathrm{Ker}((d\pi)_p)$ of vertical tangent vectors at p. Thus $\mathcal{V}$ is an integrable subbundle of $T(P)$. The diffeomorphism $L_p\colon G(\epsilon(p), \text{-}) \to P_{\pi(p)}$, given by $L_p(g) = pg^{-1}$, induces an isomorphism $dL_p\colon \mathfrak{g}_{\epsilon(p)} \to \mathcal{V}_p$, where $\mathfrak{g}_{\epsilon(p)}$ denotes the tangent space of $G(\epsilon(p), \text{-})$ at the unit $1_{\epsilon(p)}$.

Let $\mathcal{F}$ be a foliation of B. Then $\pi^*(\mathcal{F})$ is a foliation of P. An $\mathcal{F}$-*partial connection* on P is a subbundle $\mathcal{H}$ of $\pi^*(\mathcal{F}) \subset T(P)$ which satisfies the following conditions:

(i) $\pi^*(\mathcal{F}) = \mathcal{V} \oplus \mathcal{H}$,
(ii) $(d\epsilon)(\mathcal{H}) = 0$, and
(iii) $\mathcal{H}_{pg} = \mathcal{H}_p g$ for any $p \in P$ and $g \in G$ with $\epsilon(p) = \mathrm{t}(g)$.

Note that $\mathcal{H}_{pg}$ is well-defined precisely because of the condition (ii) above. The connection $\mathcal{H}$ is called *flat* if it is integrable. With a given connection $\mathcal{H}$, any tangent vector $\xi \in \pi^*(\mathcal{F})_p$ has a unique decomposition as a sum $\xi = \xi^v + \xi^h$ of its vertical and horizontal parts.

Proposition 5.38 *Let P be a principal G-bundle over B, $\mathcal{F}$ a foliation of B and $\mathcal{H}$ a flat $\mathcal{F}$-partial connection on P. Then each leaf of $\mathcal{H}$ projects by a covering projection to a leaf of $\mathcal{F}$.*

Proof Each leaf $\tilde{L}$ of $\mathcal{H}$ clearly projects by a local diffeomorphism to a leaf L of $\mathcal{F}$. Moreover, $\tilde{L}$ lies in a fibre $\epsilon^{-1}(x)$. Let $\mathrm{Iso}(\tilde{L})$ be the isotropy group of $\tilde{L}$, i.e.

$$\mathrm{Iso}(\tilde{L}) = \{g \in G(x, x) \mid \tilde{L}g = \tilde{L}\}\,.$$

The group $\mathrm{Iso}(\tilde{L})$, equipped with the discrete topology, acts freely and properly discontinuously on $\tilde{L}$, which shows that π in fact restricts to a covering projection $\tilde{L} \to L$. $\qquad\qquad\square$

6

Lie algebroids

In this final chapter we will provide a brief introduction to the theory of Lie algebroids.

Lie algebroids arise naturally as the infinitesimal parts of Lie groupoids, in complete analogy to the way that Lie algebras arise as the infinitesimal part of Lie groups. Once isolated, the concept of a Lie algebroid turns out to be a very natural one, which unifies various different types of infinitesimal structure. For example, foliated manifolds, Poisson manifolds, infinitesimal actions of Lie algebras on manifolds, and many other structures can be naturally viewed as Lie algebroids. In this way, Lie algebroids connect various themes of this book: Lie groupoids and foliations provide examples of Lie algebroids, while conversely, we will see that the basic theory of foliations which has been developed in earlier chapters can be applied to prove some of the basic structure theorems about Lie algebroids.

The plan of this chapter is as follows. In the first section, we will isolate the infinitesimal part of a given Lie groupoid, as an important way of constructing Lie algebroids. In the next section, we will introduce the abstract notion of a Lie algebroid, and present some basic examples.

The rest of this chapter is devoted to the Lie theory for Lie groupoids and Lie algebroids. The classical correspondence between (finite dimensional) Lie groups and Lie algebras is described by three 'Lie theorems'. These theorems assert that any connected Lie group can be covered by a simply connected Lie group, that maps from a simply connected Lie group into an arbitrary Lie group correspond exactly to maps between their Lie algebras, and, finally, that any Lie algebra is the Lie algebra of a Lie group. (This last, third, Lie theorem is due to E. Cartan.) Using the theory of foliations, we will show in Section 6.3 that the first part of Lie's theory can be extended to Lie algebroids. Lie's third theorem, however, fails for Lie algebroids: we will show in Section 6.4 that there are examples of Lie algebroids which are not integrable to a Lie groupoid. These examples arise naturally in the theory of transversely parallelizable foliations, discussed in Chapter 4. In fact, we will prove that any transversely parallelizable foliation on a compact manifold gives rise to a Lie algebroid, which is integrable if and only if the foliation is developable.

6.1 The Lie algebroid of a Lie groupoid

The construction of the Lie algebra $\mathfrak{g}$ of a given Lie group G extends to Lie groupoids. The infinitesimal approximation of a Lie groupoid is at the same time a natural generalization of a foliation viewed as an integrable subbundle of the tangent bundle. In this section, we shall present the construction of this infinitesimal approximation of a Lie groupoid.

Let G be a Lie groupoid. By analogy with Lie groups, we need to consider the action of G on the tangent bundle of G_1. However, this action is not everywhere defined. If $h\colon x \to y$ is an arrow in G, the composition with h gives a diffeomorphism $R_h\colon s^{-1}(y) \to s^{-1}(x)$, $R_h(g) = gh$. Therefore the natural right action of G on G_1 lifts to a right action of G on the vector bundle

$$T^s(G_1) = \operatorname{Ker}(ds) \subset T(G_1)$$

over G_1 as follows: for any $\xi \in (T^s(G_1))_g = T_g^s(G_1)$ and any $h \in G_1$ with $t(h) = s(g)$ we define

$$\xi h = dR_h(\xi) \in T_{gh}^s(G_1) \,.$$

The sections $\mathfrak{X}^s(G_1) = \Gamma(T^s(G_1))$ of this vector bundle are the vector fields on G_1 tangent to the source-fibres. In particular, $T^s(G_1)$ is involutive and $\mathfrak{X}^s(G_1)$ is a Lie subalgebra of $\mathfrak{X}(G_1)$. A *(right) G-invariant vector field* on G_1 is a vector field $X \in \mathfrak{X}^s(G_1)$ which satisfies

$$X_{gh} = X_g h$$

for any $g, h \in G_1$ with $s(g) = t(h)$. We denote by $\mathfrak{X}^s_{\mathrm{inv}}(G)$ the vector space of all right G-invariant vector fields on G_1.

Proposition 6.1 *Let G be a Lie groupoid. Then*

(i) $\mathfrak{X}^s_{\mathrm{inv}}(G)$ *is a Lie subalgebra of $\mathfrak{X}(G_1)$,*

(ii) *any G-invariant vector field on G_1 is projectable along the target map t to G_0, and*

(iii) the derivative of the target map induces a Lie algebra homomorphism $dt\colon \mathfrak{X}_{\text{inv}}^{\text{s}}(G) \to \mathfrak{X}(G_0)$.

Proof (i) Take $X, Y \in \mathfrak{X}_{\text{inv}}^{\text{s}}(G)$. We have $[X, Y] \in \mathfrak{X}^{\text{s}}(G_1)$, while for any $g, h \in G_1$ with $\text{s}(g) = \text{t}(h)$

$$[X, Y]_g h = dR_h([X, Y]_g) = [dR_h(X), dR_h(Y)]_{gh} = [X, Y]_{gh} .$$

(ii) For any arrow $g\colon x \to y \in G_1$ we have

$$dt(X_g) = dt(X_{1_y} g) = dt(dR_g(X_{1_x})) = d(\text{t} \circ R_g)(X_{1_x}) = dt(X_{1_x}) .$$

(iii) This is clear from (i) and (ii). $\qquad\qquad\square$

Any G-invariant vector field $X \in \mathfrak{X}_{\text{inv}}^{\text{s}}(G)$ is uniquely determined by its restriction (again denoted by X) to the set of units $\{1_x \,|\, x \in G_0\}$ of G, because

$$X_g = X_{1_{\text{t}(g)}} g$$

for any $g \in G$. Therefore we have an isomorphism of vector spaces

$$\mathfrak{X}_{\text{inv}}^{\text{s}}(G) \longrightarrow \Gamma(\mathfrak{g}) ,$$

where $\mathfrak{g}$ is the pull-back of the vector bundle $T^{\text{s}}(G_1)$ along the unit map uni$\colon G_0 \to G_1$ of G:

$$
\begin{array}{ccc}
\mathfrak{g} & \longrightarrow & T^{\text{s}}(G_1) \\
\pi \downarrow & & \downarrow \pi \\
G_0 & \xrightarrow{\ \text{uni}\ } & G_1
\end{array}
$$

Thus there is a unique Lie algebra structure on $\Gamma(\mathfrak{g})$ for which the isomorphism of vector spaces $\mathfrak{X}_{\text{inv}}^{\text{s}}(G) \cong \Gamma(\mathfrak{g})$ is an isomorphism of Lie algebras. If X is a section of $\Gamma(\mathfrak{g})$, we shall denote the unique G-invariant extension of X to G_1 again by X, and sometimes also by XG to avoid possible ambiguities. With this notation, we have $[X, Y]G = [XG, YG]$ and $(fX)G = (f \circ \text{t})XG$ for any $f \in C^\infty(G_0)$.

The derivative of the target map restricts to an 'anchor' homomorphism an$\colon \mathfrak{g} \to T(G_0)$ of vector bundles over G_0 by

$$\text{an}(\xi) = dt(\xi) .$$

This map induces a homomorphism of Lie algebras

$$\Gamma(\text{an})\colon \Gamma(\mathfrak{g}) \longrightarrow \mathfrak{X}(G_0) ,$$

which corresponds to the map $dt\colon \mathfrak{X}^{\mathrm{s}}_{\mathrm{inv}}(G) \to \mathfrak{X}(G_0)$ of Proposition 6.1 (iii) under the isomorphism $\mathfrak{X}^{\mathrm{s}}_{\mathrm{inv}}(G) \cong \Gamma(\mathfrak{g})$. Thus

$$\Gamma(\mathrm{an})(X) = dt(XG)$$

for any $X \in \Gamma(\mathfrak{g})$.

Now take any $X, Y \in \Gamma(\mathfrak{g})$ and $f \in C^\infty(G_0)$. The Leibniz identity for vector fields implies

$$
\begin{aligned}
[X, fY]G &= [XG, (fY)G] \\
&= [XG, (f \circ \mathrm{t})YG] \\
&= (f \circ \mathrm{t})[XG, YG] + (XG)(f \circ \mathrm{t})YG \\
&= (f \circ \mathrm{t})[X, Y]G + (dt(XG)(f) \circ \mathrm{t})YG \\
&= (f \circ \mathrm{t})[X, Y]G + (dt(XG)(f)Y)G \\
&= (f[X, Y])G + (\Gamma(\mathrm{an})(X)(f)Y)G \; .
\end{aligned}
$$

Therefore the relation between the Lie algebra and the $C^\infty(G_0)$-module structures on $\Gamma(\mathfrak{g})$ is

$$[X, fY] = f[X, Y] + \Gamma(\mathrm{an})(X)(f)Y \; .$$

The vector bundle $\mathfrak{g}$ over G_0 with the structure described above is called the associated *Lie algebroid of the Lie groupoid* G. In the next section we shall give an abstract definition of a Lie algebroid, but first let us compute this Lie algebroid for some special Lie groupoids we already know.

Examples 6.2 (1) If G is a Lie group, viewed as a Lie groupoid over a one point space, the associated bundle $\mathfrak{g}$ is a bundle over this one point space with a Lie algebra structure on its sections. This Lie algebra is precisely the Lie algebra of right invariant vector fields on G, isomorphic to the tangent space of G at the unit of the group.

(2) Let G be the holonomy groupoid $\mathrm{Hol}(M, \mathcal{F})$ of a foliation $\mathcal{F}$ of a manifold M, and let $\mathfrak{g}$ be its Lie algebroid. For any $x \in M$, the source-fibre $\mathrm{Hol}(M, \mathcal{F})(x, \text{-})$ is the holonomy cover of the leaf L through x via the target map, therefore the anchor map of $\mathfrak{g}$ maps $\mathfrak{g}_x$ bijectively onto the subspace $T(\mathcal{F})_x$ of vectors in $T_x(M)$ tangent to the leaf L. Thus we may identify $\mathfrak{g}$ with the foliation $\mathcal{F}$ itself viewed as a subbundle $T(\mathcal{F})$ of $T(M)$.

Note that by the same argument, the Lie algebroid of the monodromy groupoid $\mathrm{Mon}(M, \mathcal{F})$ can also be identified with $T(\mathcal{F})$.

6.2 Definition and examples of Lie algebroids

In this section we give the abstract definition of a Lie algebroid as well as some of the main examples.

Let M be a manifold. A *Lie algebroid* over M is a vector bundle $\pi\colon \mathfrak{g} \to M$ over M, together with a map an$\colon \mathfrak{g} \to T(M)$ of vector bundles over M and a (real) Lie algebra structure $[\,\text{-}\,,\,\text{-}\,]$ on the vector space $\Gamma\mathfrak{g}$ of sections of $\mathfrak{g}$, such that

(i) the induced map $\Gamma(\text{an})\colon \Gamma\mathfrak{g} \to \mathfrak{X}(M)$ is a Lie algebra homomorphism, and

(ii) the *Leibniz identity*

$$[X, fY] = f[X, Y] + \Gamma(\text{an})(X)(f)Y$$

holds for any $X, Y \in \Gamma\mathfrak{g}$ and any $f \in C^\infty(M)$.

The map an is called the *anchor* of the Lie algebroid $\mathfrak{g}$. The map $\Gamma(\text{an})$ is often simply denoted by 'an' as well, and also called the anchor. The manifold M is called the *base manifold* of the Lie algebroid $\mathfrak{g}$.

Let $\mathfrak{g}$ and $\mathfrak{h}$ be two Lie algebroids over the same base manifold M. A morphism of vector bundles $\mathfrak{g} \to \mathfrak{h}$ over M is a *morphism of Lie algebroids* if it commutes with the anchors and preserves the Lie algebra structure on sections.

It is a bit more complicated to define a morphism of Lie algebroids over different base manifolds. Let $\mathfrak{g}$ be a Lie algebroid over M and $\mathfrak{h}$ a Lie groupoid over N. A bundle map $\Phi\colon \mathfrak{h} \to \mathfrak{g}$ over $\phi\colon N \to M$ is a *morphism of Lie algebroids* if an $\circ\, \Phi = d\phi \circ$ an (i.e. it preserves the anchor) and if it preserves the Lie bracket in the following sense. First, note that Φ can be equivalently viewed as a bundle map $(\pi, \Phi)\colon \mathfrak{h} \to \phi^*\mathfrak{g} = N \times_M \mathfrak{g}$ over N, and recall that the map $C^\infty(N) \otimes_{C^\infty(M)} \Gamma\mathfrak{g} \to \Gamma(\phi^*\mathfrak{g})$, which sends $f \otimes X$ to $f\phi^*(X) = f(\text{id}, X \circ \phi)$, is an isomorphism of $C^\infty(N)$-modules (Greub–Halperin–Vanstone (1978), page 83):

$$
\begin{array}{ccc}
\mathfrak{h} \xrightarrow{\;(\pi,\Phi)\;} \phi^*\mathfrak{g} & \longrightarrow & \mathfrak{g} \\
\pi \downarrow \quad \phi^*(X) \Updownarrow & \quad X \Updownarrow \\
N = N \xrightarrow{\;\phi\;} M
\end{array}
$$

Now the bundle map Φ preserves the bracket if for any $Y, Y' \in \Gamma(\mathfrak{h})$, with $(\pi, \Phi) \circ Y = \sum_i f_i \phi^*(X_i)$ and $(\pi, \Phi) \circ Y' = \sum_j f'_j \phi^*(X'_j)$, one has

$$(\pi, \Phi) \circ [Y, Y'] = \sum_{i,j} f_i f'_j \phi^*([X_i, X'_j]) + \sum_j \mathrm{an}(Y)(f'_j) \phi^*(X'_j)$$
$$- \sum_i \mathrm{an}(Y')(f_i) \phi^*(X_i) .$$

Equivalently, the map Φ preserves the bracket if

$$(\mathrm{an}, \Phi) \colon \Gamma(\mathfrak{h}) \longrightarrow K \subset \mathfrak{X}(N) \oplus \Gamma(\phi^* \mathfrak{g})$$

is a homomorphism of Lie algebras, where K is the kernel of the linear map $\kappa \colon \mathfrak{X}(N) \oplus \Gamma(\phi^* \mathfrak{g}) \to \Gamma(\phi^* T(M)) = \Gamma(N \times_M T(M))$,

$$\kappa \left(Z \oplus \sum_i f_i \phi^*(X_i) \right) = (\mathrm{id}, d\phi \circ Z) - \sum_i f_i \phi^*(\mathrm{an}(X_i)) .$$

Here K has the structure of a Lie algebra, in which the Lie bracket of $Z \oplus \sum_i f_i \phi^*(X_i)$ and $Z' \oplus \sum_j f'_j \phi^*(X'_j)$ has $[Z, Z']$ as its first component and

$$\sum_{i,j} f_i f'_j \phi^*([X_i, X'_j]) + \sum_j Z(f'_j) \phi^*(X'_j) - \sum_i Z'(f_i) \phi^*(X_i)$$

as its second.

A generic example of a Lie algebroid over M is of course the Lie algebroid of a Lie groupoid with M as the space of objects, as described in Section 6.1. One easily checks that the differential of a homomorphism $H \to G$ of Lie groupoids induces a morphism of the associated Lie algebroids in a functorial way.

A Lie algebroid $\mathfrak{g}$ is called *integrable* if it is isomorphic to the Lie algebroid associated to a Lie groupoid G. If this is the case, then G is called an *integral* of $\mathfrak{g}$.

Every finite dimensional Lie algebra is a Lie algebroid over a one point space, and by Lie's third theorem it is integrable. However, there exist Lie algebroids which are not integrable. The first example of such a Lie algebroid was given by Almeida and Molino (1985), and we shall present their construction in Section 6.4.

Examples 6.3 (1) Any manifold M can be viewed as a Lie algebroid in two ways, by taking the zero bundle over M (which we shall denote simply by M), or by taking the tangent bundle over M with the identity map for the anchor (we shall denote this Lie algebroid by $T(M)$). Both these Lie algebroids are integrable, the first by the unit groupoid on M, and the second by the pair groupoid over M.

(2) Any vector bundle E over M can be viewed as a Lie algebroid over M, with zero bracket and anchor. More generally, a vector bundle E over M with a smoothly varying Lie algebra structure on its fibres (i.e. a bundle of Lie algebras) can be viewed as a Lie algebroid over M with zero anchor. Any bundle of Lie algebras is integrable by a bundle of Lie groups (which may not be locally trivial nor Hausdorff), by a result of Douady and Lazard (1966).

(3) A foliation $\mathcal{F}$ of M is given by an involutive subbundle $T(\mathcal{F})$ of $T(M)$. Thus a foliation of M is (up to an isomorphism of Lie algebroids over M) the same thing as a Lie algebroid over M with injective anchor map. Any foliation $\mathcal{F}$ on M is integrable as a Lie algebroid by $\mathrm{Mon}(M, \mathcal{F})$ and also by $\mathrm{Hol}(M, \mathcal{F})$ (Example 6.2 (2)).

(4) Let M be a manifold equipped with an *infinitesimal action* of a Lie algebra $\mathfrak{g}$, i.e. a Lie algebra homomorphism $\gamma \colon \mathfrak{g} \to \mathfrak{X}(M)$. The trivial bundle $\mathfrak{g} \times M$ over M has the structure of a Lie algebroid, with anchor given by $\mathrm{an}(\xi, x) = \gamma(\xi)_x$, and Lie bracket

$$[u, v](x) = [u(x), v(x)] + (\gamma(u(x))(v))(x) - (\gamma(v(x))(u))(x) ,$$

for $u, v \in C^\infty(M, \mathfrak{g}) \cong \Gamma(M, \mathfrak{g} \times M)$ and $x \in M$. (This is the unique way to define a bracket which satisfies the Leibniz rule and agrees with the bracket of $\mathfrak{g}$ for constant functions u and v.) This Lie algebroid is called the *transformation Lie algebroid* associated to the infinitesimal action, and denoted by $\mathfrak{g} \ltimes M$.

The derivative of an action of a Lie group G on M gives an infinitesimal action of the Lie algebra $\mathfrak{g}$ of G on M, and the associated Lie algebroid is integrable by the associated action groupoid $G \ltimes M$. However, it is easy to construct infinitesimal actions of $\mathfrak{g}$ on M which do not come from an action of G on M. Nevertheless, the associated Lie algebroid of any such action is integrable. The following construction of the integral groupoid is due to Dazord (1997).

Let G be a Lie group with the given Lie algebra $\mathfrak{g}$ and let $\gamma \colon \mathfrak{g} \to \mathfrak{X}(M)$ be an infinitesimal action of $\mathfrak{g}$ on a manifold M. Now consider the foliation $\mathcal{F}$ on $M \times G$ given at any $(x, g) \in M \times G$ by

$$T(\mathcal{F})_{(x,g)} = \{(\gamma(\xi)_x, \xi g) | \xi \in \mathfrak{g}\} .$$

Here $\xi g \in T_g(G)$ denotes the right translation of $\xi \in T_e(G)$. The action of G on itself by right translations gives an action on $M \times G$ which leaves the foliation $\mathcal{F}$ invariant. The group G therefore acts also on the homotopy classes of paths $\mathrm{Mon}(M \times G, \mathcal{F})_1$. Now Lemma 5.9 implies that

$$H = \mathrm{Mon}(M \times G, \mathcal{F})/G$$

is a Lie groupoid over $(H \times G)/G = M$, and the reader can easily verify that H integrates the Lie algebroid $\mathfrak{g} \ltimes M$ associated to the infinitesimal action γ.

(5) Let (M, Π) be a Poisson manifold. There is a natural Lie algebra structure on $\Omega^1(M)$ which makes $T^*(M)$ into a Lie algebroid over M. The anchor of this Lie algebroid is $-\tilde{\Pi}$, where $\tilde{\Pi} \colon T^*(M) \to T(M)$ is induced by the bivector field Π. For details, see e.g. Cannas da Silva–Weinstein (1999).

(6) A Lie algebroid $\mathfrak{g}$ over M is said to be *regular* if its anchor map $\mathrm{an} \colon \mathfrak{g} \to T(M)$ has constant rank. In this case, the image and the kernel of the anchor are subbundles of $T(M)$ and of $\mathfrak{g}$ respectively, and furthermore, the image of the anchor map is the tangent bundle $T(\mathcal{F})$ of a foliation $\mathcal{F}$ of M, while the kernel $\mathrm{Ker}(\mathrm{an})$ is a bundle of Lie algebras. We thus have a short exact sequence of Lie algebroids over M

$$0 \longrightarrow \mathrm{Ker}(\mathrm{an}) \longrightarrow \mathfrak{g} \stackrel{\mathrm{an}}{\longrightarrow} T(\mathcal{F}) \longrightarrow 0$$

and the associated short exact sequence of Lie algebras of sections,

$$0 \longrightarrow \Gamma(\mathrm{Ker}(\mathrm{an})) \longrightarrow \Gamma(\mathfrak{g}) \stackrel{\mathrm{an}}{\longrightarrow} \mathfrak{X}(\mathcal{F}) \longrightarrow 0 .$$

The Lie algebroid $\mathfrak{g}$ is called *transitive* if it has surjective anchor. In this case we have $T(\mathcal{F}) = T(M)$, and the associated short exact sequence of Lie algebras (and $C^\infty(M)$-modules)

$$0 \longrightarrow \Gamma(\mathrm{Ker}(\mathrm{an})) \longrightarrow \Gamma(\mathfrak{g}) \stackrel{\mathrm{an}}{\longrightarrow} \mathfrak{X}(M) \longrightarrow 0$$

is also called an (abstract) *Atiyah sequence*.

Note that the Lie algebroid $\mathfrak{g}$ of a Lie groupoid G over a connected base G_0 is transitive if and only if the Lie groupoid G is transitive, by Proposition 5.14 (iii). Since any transitive Lie groupoid is in fact a gauge groupoid of a principle G-bundle $\pi \colon P \to M$ for a Lie group G by Proposition 5.14 (v), we can construct the associated Lie algebroid $\mathfrak{g}$ directly from P (and this is in fact the construction given by Atiyah). To do this, observe that G acts on $T(P)$, and that Lemma 5.35 implies that $T(P)/G$ is a manifold making the induced map $T(P)/G \to P/G = M$

into a vector bundle:

$$
\begin{array}{ccc}
T(P) & \longrightarrow & T(P)/G \\
\pi \downarrow & & \downarrow \\
P & \longrightarrow & P/G = M
\end{array}
$$

The space of sections of $T(P)/G$ can now be identified with the Lie algebra $\mathfrak{X}_G(P)$ of G-invariant vector fields on P, while the anchor map an: $T(P)/G \to T(M)$ of $T(P)/G$ is induced by $d\pi \colon T(P) \to T(M)$. This makes $T(P)/G$ into a transitive Lie algebroid over M isomorphic to the Lie algebroid of the gauge groupoid Gauge(P) of P. We may conclude:

Corollary 6.4 *A transitive Lie algebroid over a connected manifold M is integrable if and only if it is isomorphic to the Lie algebroid $T(P)/G$ of a principal G-bundle P over M, for a Lie group G.*

6.3 Lie theory for Lie groupoids

With the exception of Lie's third theorem, the classical Lie theory for Lie groups and Lie algebras extends to Lie groupoids and Lie algebroids. In this section, we shall give a short presentation of the main results.

The Lie theory for groups states that every finite dimensional Lie algebra is the Lie algebra of a unique connected and simply connected Lie group. Something similar holds for *integrable* Lie algebroids.

A Lie groupoid G is *source-connected* if all the fibres of the source map s: $G_1 \to G_0$ are connected. Furthermore, the Lie algebroid G is *source-simply-connected* if each such fibre is connected and simply connected.

Example 6.5 The monodromy groupoid Mon($M, \mathcal{F}$) of a foliated manifold is source-simply-connected.

Proposition 6.6 *For any Lie groupoid G there exist a source-simply-connected Lie groupoid $\tilde{G}$ over G_0 and a morphism of Lie groupoids $\tilde{G} \to G$ over G_0 which induces an isomorphism of the associated Lie algebroids.*

REMARK. The covering Lie groupoid $\tilde{G}$ of G is unique up to an isomorphism, by Proposition 6.8 below.

Proof (of Proposition 6.6) Let $\mathcal{F}$ be the foliation of G_1 given by the fibres of the source map, and let $\mathrm{Mon}(G_1, \mathcal{F})$ be its monodromy groupoid over G_1. The multiplication of the Lie groupoid G turns the space G_1 into a principal right G-bundle over G_0, and this principal action preserves the foliation $\mathcal{F}$. Thus G also acts on the monodromy groupoid of $\mathcal{F}$. By Lemma 5.9 we obtain the quotient Lie groupoid

$$\tilde{G} = \mathrm{Mon}(G_1, \mathcal{F})/G \,,$$

which is a groupoid over G_0. Since any monodromy groupoid is source-simply-connected and $\mathrm{Mon}(G_1, \mathcal{F})$ has the same source-fibres as its quotient by G (by the Remark after the proof of Lemma 5.9), the Lie groupoid $\tilde{G}$ is again source-simply-connected. The morphism of Lie groupoids

$$\phi \colon \mathrm{Mon}(G_1, \mathcal{F}) \longrightarrow G$$

given by t: $G_1 \to G_0$ on objects and by $\phi(\sigma) = \sigma(1)\sigma(0)^{-1}$ on arrows (here σ denotes the homotopy class of a path inside a leaf of $\mathcal{F}$) factors to give the required map $\tilde{G} \to G$.   □

Let $\mathfrak{g}$ be a Lie algebroid over M, and let N be an immersed submanifold of M. A *Lie subalgebroid* of $\mathfrak{g}$ over N is a subbundle $\mathfrak{h}$ of the restriction $\mathfrak{g}|_N$, equipped with a Lie algebroid structure such that the inclusion $\mathfrak{h} \to \mathfrak{g}$ is a morphism of Lie algebroids.

The same methods involved in the construction of the source-simply-connected Lie groupoid $\tilde{G}$ can be used to prove that a Lie subalgebroid of an integrable Lie algebroid is again integrable.

Proposition 6.7 *Any Lie subalgebroid of an integrable Lie algebroid is integrable.*

Proof Let $\mathfrak{g}$ be the Lie algebroid of a Lie groupoid G, and let $\mathfrak{h}$ be a Lie subalgebroid of $\mathfrak{h}$ over $H_0 \subset G_0$ of $\mathfrak{g}$. Denote by $I \colon \mathfrak{h} \to \mathfrak{g}$ the inclusion of Lie algebroids over the injective immersion $\iota \colon H_0 \to G_0$, and let

$$M = H_0 \times_{G_0} G_1$$

denote the pull-back of t: $G_1 \to G_0$ along ι. Consider the foliation $\mathcal{F}$ of M given by

$$T(\mathcal{F})_{(y,g)} = \{(\mathrm{an}(\zeta), I(\zeta)g) \mid \zeta \in \mathfrak{h}_y\} \,.$$

The composition of the Lie groupoid G gives M the structure of a right principal G-bundle over $\tilde{H}_0$, and the foliation $\mathcal{F}$ is invariant under the

G-action. Thus the monodromy groupoid $\mathrm{Mon}(M, \mathcal{F})$ also carries a right G-action, and by Lemma 5.9 we obtain a quotient Lie groupoid

$$H = \mathrm{Mon}(M, \mathcal{F})/G$$

over H_0. This Lie groupoid H integrates $\mathfrak{h}$. □

REMARK. By the integrability of morphisms between integrable Lie algebroids (Proposition 6.8 below), there is a map $H \to G$ which is in fact an immersion.

Finally we shall give a quick proof of the fact that any morphism of integrable Lie algebroids can be integrated to a unique morphism of the integral Lie groupoids, provided that the domain Lie groupoid is source-simply-connected.

Proposition 6.8 *Let G and H be Lie groupoids, with H source-simply-connected, and let $\Phi\colon \mathfrak{h} \to \mathfrak{g}$ be a morphism of their Lie algebroids over $\phi\colon H_0 \to G_0$. Then ϕ can be extended to a unique morphism of Lie groupoids $H \to G$ which integrates Φ.*

Proof Let $P = H_1 \times_{G_0} G_1$ be the pull-back of $\mathrm{t}\colon G_1 \to G_0$ along the map $\phi \circ \mathrm{t}\colon H_1 \to G_0$. Thus P is a (trivial) principal G-bundle over H_1, with the obvious right action with respect to the map $\epsilon = \mathrm{s} \circ \mathrm{pr}_2$. Let $\mathcal{F}$ be the foliation of H_1 by the source-fibres. Define a partial connection $\mathcal{H}$ on P by

$$\mathcal{H}_{(h,g)} = \{(\zeta h, \Phi(\zeta)g) \mid \zeta \in \mathfrak{h}_{\mathrm{t}(h)}\} \, .$$

This is a flat connection on P because Φ preserves the bracket. Now take any $y \in H_0$, and denote by $\tilde{L}_y$ the leaf of $\mathcal{H}$ through the point $(1_y, 1_{\phi(y)})$. By Proposition 5.38, $\tilde{L}_y$ is a connected covering space over the corresponding leaf of $\mathcal{F}$, i.e. the source-fibre $\mathrm{s}^{-1}(y)$. Since the source-fibres of H are simply connected, the projection $\tilde{L}_y \to \mathrm{s}^{-1}(y)$ must be a diffeomorphism. Denote by ν_y the inverse of this diffeomorphism.

The union of the maps ν_y gives us a map $\nu\colon H_1 \to P$. Observe that this map is smooth. Indeed, it is the extension by holonomy of its restriction $H_0 \to P$, which is the smooth map sending $y \in H_0$ to $(1_y, 1_{\phi(y)})$. Then we define $\phi\colon H_1 \to G_1$ to be the composition

$$\phi = \mathrm{pr}_2 \circ \nu\colon H_1 \longrightarrow G_1 \, .$$

In particular, ϕ maps $\mathrm{s}^{-1}(y)$ to $\mathrm{s}^{-1}(\phi(y))$. It is easy to see that ϕ

is a morphism of Lie groupoids $H \to G$. For any $\zeta \in \mathfrak{h}_y$ we have $d(\phi)(\zeta) = d(\mathrm{pr}_2)(\zeta, \Phi(\zeta)) = \Phi(\zeta)$, so ϕ integrates Φ. $\square$

6.4 Integrability and developable foliations

In this section we shall describe the Lie algebroid of a transversely parallelizable foliation, and prove the theorem of Almeida and Molino which states that this Lie algebroid is integrable if and only if the foliation is developable. In particular, this gives concrete examples of non-integrable Lie algebroids.

Let $\mathcal{F}$ be a transversely parallelizable foliation of codimension q of a compact connected manifold M. Recall from Section 4.1 that $\mathcal{F}$ is homogeneous and hence contained in the associated basic foliation $\mathcal{F}_{\mathrm{bas}}$, the leaves of which are the fibres of a submersion $\pi_{\mathrm{bas}} \colon M \to W$ into a Hausdorff manifold $W = M/\mathcal{F}_{\mathrm{bas}}$. This submersion induces isomorphisms $\Omega^0_{\mathrm{bas}}(M, \mathcal{F}) \cong C^\infty(W)$ and $l(M, \mathcal{F}_{\mathrm{bas}}) \cong \mathfrak{X}(W)$, while the inclusion $\mathcal{F} \subset \mathcal{F}_{\mathrm{bas}}$ gives a diagram of $C^\infty(W)$-linear Lie algebra homomorphisms (Lemma 4.5) with exact rows:

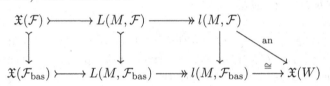

The Lie algebra $l(M, \mathcal{F})$ is a free $C^\infty(W)$-module of rank q, hence can be viewed as the space of sections of the trivial vector bundle $E = \mathbb{R}^q \times W$ over W (the explicit isomorphism between $l(M, \mathcal{F})$ and $\Gamma(E)$ depends on the choice of a transverse parallelism on $(M, \mathcal{F})$). Since the map $l(M, \mathcal{F}) \to \mathfrak{X}(W)$ on the right of the diagram is $C^\infty(W)$-linear, it defines a map of vector bundles an$\colon E \to T(W)$ over W. The Lie bracket of $l(M, \mathcal{F})$ defines a Lie bracket on $\Gamma(E)$.

Lemma 6.9 *The Lie bracket on $\Gamma(E)$, together with the anchor map* an$\colon E \to T(W)$ *described above, gives E the structure of a transitive Lie algebroid over W.*

Proof For any $Y \in L(M, \mathcal{F})$ and its class $\bar{Y} \in l(M, \mathcal{F})$, the $C^\infty(W)$-module structure on $l(M, \mathcal{F})$ is given by $f\bar{Y} = \overline{(f \circ \pi_{\mathrm{bas}})Y}$. Now the Leibniz identity for E follows from the Leibniz identity for vector fields on W and the fact that $l(M, \mathcal{F}) \to \mathfrak{X}(W)$ is a $C^\infty(W)$-linear Lie algebra homomorphism. This proves that we indeed have a Lie algebroid. To see that it is transitive, we need to show that an: $E \to T(W)$ is surjective. Take any tangent vector $\xi \in T_w(W)$ at a point $w \in W$. Choose a point $x \in M$ with $\pi_{\text{bas}}(x) = w$ and a tangent vector $\zeta \in T_x(M)$ with $d\pi_{\text{bas}}(\zeta) = \xi$. For a transverse parallelism $\bar{Y}_1, \ldots, \bar{Y}_q$ on $(M, \mathcal{F})$, the vectors $(Y_1)_x, \ldots, (Y_q)_x$ span a subspace of $T_x(M)$ complementary to $T_x(\mathcal{F})$, hence we can choose $a_1, \ldots, a_q \in \mathbb{R}$ such that

$$a_1(Y_1)_x + \cdots + a_q(Y_q)_x - \zeta \in T_x(\mathcal{F}) \ .$$

This yields that $Y = a_1 Y_1 + \cdots + a_q Y_q \in L(M, \mathcal{F})$ with $\text{an}(\bar{Y}_x) = \xi$. $\quad\square$

We shall call the Lie algebroid E over W constructed above the *basic Lie algebroid* of the foliation $(M, \mathcal{F})$, and denote it by

$$\mathfrak{b}(M, \mathcal{F}) \ .$$

Note that this Lie algebroid, which is defined for any transversely parallelizable foliation of a compact connected manifold, is independent, up to isomorphism, of the choice of a transverse parallelism on $(M, \mathcal{F})$ (which we used to identify $l(M, \mathcal{F})$ with $\Gamma(E)$).

Recall that a foliation $(M, \mathcal{F})$ is *developable* if its lifted foliation $\tilde{\mathcal{F}}$ to a covering space $\tilde{M}$ of M is strictly simple (Example 1.1 (2)).

Theorem 6.10 (Almeida–Molino) *A transversely parallelizable foliation of a compact connected manifold is developable if and only if its basic Lie algebroid is integrable.*

Proof Let $\mathcal{F}$ be a transversely parallelizable foliation of a compact manifold M, let $\mathcal{F}_{\text{bas}}$ be the basic foliation of $\mathcal{F}$, and let $\pi_{\text{bas}} \colon M \to W = M/\mathcal{F}_{\text{bas}}$ be the associated basic fibre bundle, all as above.

($\Leftarrow$) Suppose that the basic Lie algebroid $\mathfrak{b}(M, \mathcal{F})$ of $\mathcal{F}$ is integrable. Since this Lie algebroid is transitive, Corollary 6.4 provides a Lie group G, a principal G-bundle $\pi \colon P \to W$ over the basic manifold W of $\mathcal{F}$ and an isomorphism of Lie algebroids

$$\Phi \colon \mathfrak{b}(M, \mathcal{F}) \longrightarrow T(P)/G \ ,$$

which induces an isomorphism of Lie algebras $\Phi \colon l(M, \mathcal{F}) \to \mathfrak{X}_G(P)$. Now let $S = M \times_W P$ be the principal G-bundle over M obtained as the pull-back of P along π_{bas}:

$$
\begin{array}{ccc}
S & \xrightarrow{\ \text{pr}_2\ } & P \\
\text{pr}_1 \downarrow & & \downarrow \\
M & \xrightarrow{\ \pi_{\text{bas}}\ } & W
\end{array}
$$

For any $Y \in L(M, \mathcal{F})$ we have $\tilde{Y} = (Y, \Phi(\bar{Y})) \in \mathfrak{X}_G(S)$. We define a foliation $\mathcal{G}$ of S by

$$
T(\mathcal{G})_{(x,p)} = \{\tilde{Y}_{(x,p)} \mid Y \in L(M, \mathcal{F})\} \ .
$$

This is indeed a subbundle of $T(S)$ of rank $n = \dim M$, which is involutive because Φ preserves the bracket. The foliation $\mathcal{G}$ is also invariant under the action of G, hence it is a flat connection on the principal G-bundle S over M. Let $\tilde{M}$ be any leaf of the foliation $\mathcal{G}$. The projection pr_1 restricts to a covering projection $\tilde{M} \to M$. Let $\tilde{\mathcal{F}}$ be the lift of the foliation $\mathcal{F}$ to $\tilde{M}$:

$$
\begin{array}{ccc}
(\tilde{M}, \tilde{\mathcal{F}}) & \xrightarrow{\ \text{pr}_2\ } & P \\
\text{pr}_1 \downarrow & & \\
(M, \mathcal{F}) & &
\end{array}
$$

Note that $\text{pr}_2 \colon \tilde{M} \to P$ is a submersion, because

$$
\{\Phi(\bar{Y})_p \mid Y \in L(M, \mathcal{F})\} = T_p(P)
$$

for any $p \in P$. Clearly $T(\tilde{\mathcal{F}})$ is a subbundle of $\text{Ker}(d(\text{pr}_2)) \colon \tilde{M} \to P$; comparing the dimensions

$$
\text{codim}\, \tilde{\mathcal{F}} = \text{codim}\, \mathcal{F} = \text{rank}(T(P)/G) = \dim P
$$

we conclude that $T(\tilde{\mathcal{F}}) = \text{Ker}(d(\text{pr}_2)) \colon \tilde{M} \to P$. Thus $\tilde{\mathcal{F}}$ is simple, given by the submersion $\tilde{M} \to P$.

The foliation $\mathcal{F}$ is homogeneous by Theorem 4.8, so the lift $\tilde{\mathcal{F}}$ is homogeneous as well. Thus $\tilde{\mathcal{F}}$ is strictly simple by Theorem 4.3 (vi).

($\Rightarrow$) Now suppose that $\mathcal{F}$ is developable. Thus, there is a covering space $u \colon \tilde{M} \to M$ such that the pull-back $\tilde{\mathcal{F}} = u^*(\mathcal{F})$ is a strictly simple foliation of $\tilde{M}$. Let $\psi \colon \tilde{M} \to P$ be a surjective submersion with connected fibres defining $\tilde{\mathcal{F}}$. Note that, by replacing P with its universal cover and ψ by the pull-back along the universal cover of P, we may assume that P is simply connected.

The covering projection $u\colon \tilde{M} \to M$ induces a surjective submersion $\pi\colon P \to W$ making the square

$$
\begin{array}{ccc}
(\tilde{M}, \tilde{\mathcal{F}}) & \xrightarrow{\ \psi\ } & P \\
{\scriptstyle u}\downarrow & & \downarrow{\scriptstyle \pi} \\
(M, \mathcal{F}) & \xrightarrow{\ \pi_{\mathrm{bas}}\ } & W
\end{array}
$$

commute. Consider the following diagram of $C^\infty(W)$-linear Lie algebra homomorphisms with exact rows, the upper half of which is given by the pull-back along u, while the lower half is identical to the diagram just above Lemma 6.9:

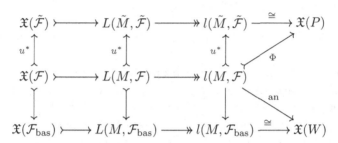

Write Φ for the indicated composition. Note that the map Φ is a monomorphism because the upper left square is a pull-back. We shall write $\mathfrak{l} = \Phi(l(M,\mathcal{F})) \subset \mathfrak{X}(P)$ for the image of Φ. Note that the vector fields in $\mathfrak{l}$ are projectable along π to W (the projection being given by an $\circ\, \Phi^{-1}$). It is now sufficient to prove that the map $\pi\colon P \to W$ can be equipped with the structure of a principal G-bundle for a Lie group G, in such a way that $\mathfrak{l} \subset \mathfrak{X}(P)$ becomes the subalgebra $\mathfrak{X}_G(P)$ of G-invariant vector fields on P.

To this end, consider the fibred product $R = P \times_W P$, and define a 'diagonal' foliation $\mathcal{R}$ of R by

$$
T(\mathcal{R})_{(p,p')} = \{(\Phi(\bar{Y})_p, \Phi(\bar{Y})_{p'}) \mid Y \in L(M,\mathcal{F})\} \subset T_{(p,p')}(R)\ .
$$

Notice that, since M is compact, any vector field $Y \in L(M,\mathcal{F})$ is complete, and hence so are its pull-back $u^*(Y)$ to $\tilde{M}$ and the image $\Phi(\bar{Y}) \in \mathfrak{X}(P)$ of its class $\bar{Y} \in l(M,\mathcal{F})$, as is its image $\mathrm{an}(\bar{Y}) \in \mathfrak{X}(W)$ (because W is also compact).

Now let w be a point of W, and choose $x \in M$ with $\pi_{\mathrm{bas}}(x) = w$. Choose projectable vector fields $Y_1, \ldots, Y_{q'} \in L(M,\mathcal{F})$ such that $(Y_1)_x, \ldots, (Y_{q'})_x$ form a basis of a subspace of $T_x(M)$ complementary to $T_x(\mathcal{F}_{bas})$ (thus $q' = \operatorname{codim}\mathcal{F} = \dim W$). Then $\operatorname{an}(\bar{Y}_1),\ldots,\operatorname{an}(\bar{Y}_{q'})$ form a frame of $T(W)$ near w. The map $S\colon \mathbb{R}^{q'} \to W$ given by

$$S(t) = (e^{t_1 \operatorname{an}(\bar{Y}_1)} \circ \cdots \circ e^{t_{q'} \operatorname{an}(\bar{Y}_{q'})})(w)\,, \qquad t = (t_1,\ldots,t_{q'})\,,$$

is a diffeomorphism on a neighbourhood U of $0 \in \mathbb{R}^{q'}$ and hence defines a local chart on W around w, while the map $T\colon \pi^{-1}(w) \times U \to P$ given by

$$T(p,t) = (e^{t_1 \Phi(\bar{Y}_1)} \circ \cdots \circ e^{t_{q'} \Phi(\bar{Y}_{q'})})(p)\,, \qquad t = (t_1,\ldots,t_{q'})\,,$$

defines a local trivialization of P as a fibre bundle over the chart $S(U)$:

$$
\begin{array}{ccc}
\pi^{-1}(w) \times U & \xrightarrow{\ T\ } & P \\
{\scriptstyle \mathrm{pr}_2}\downarrow & & \downarrow{\scriptstyle \pi} \\
U & \xrightarrow{\ S\ } & W
\end{array}
$$

Consider the leaf $L_{p,p'}$ of $\mathcal{R}$ through a point $(p,p') \in R$. Let $\bar{Y}_1,\ldots,\bar{Y}_q$ be a transverse parallelism of $(M,\mathcal{F})$. Using the flows of the complete vector fields $\Phi(\bar{y}_1),\ldots\Phi(\bar{Y}_q)$ on P, and the associated flows of the 'diagonal' vector fields $(\Phi(\bar{Y}_i),\Phi(\bar{Y}_i))$ on R, we can show in a similar way that the two projections $\mathrm{pr}_1\colon L_{p,p'} \to P$ and $\mathrm{pr}_2\colon L_{p,p'} \to P$ are coverings, hence diffeomorphisms because P is simply connected. Therefore for any $(p,p') \in R$ we obtain a diffeomorphism $\tau_{p,p'}\colon P \to P$ as the combination of these two projections, as in the following diagram:

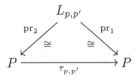

Let G be the group of all diffeomorphisms of P of this form,

$$G = \{\tau_{p,p'} \mid (p,p') \in R\}\,.$$

This group clearly acts freely and transitively along the fibres of the map $\pi\colon P \to W$. By fixing a point $p \in P$ we can identify G with the fibre $P_{\pi(p)}$, which gives a smooth structure on G. With this, G becomes a Lie group and P a principal G-bundle. $\qquad\square$

Exercise 6.11 To construct a transversely parallelizable foliation of a compact manifold which is not developable, take a compact simply connected Lie group G with a non-closed subgroup H. Show that the right cosets of H form a transversely parallelizable foliation of G. This foliation is not developable because G is simply connected and the leaves of this foliation are not closed. Hence, the associated basic Lie algebroid of this foliation is not integrable.

References and further reading

Given the introductory nature of this book, we have tried to keep the bibliography short. There are many introductions to the theory of foliations, for example Lawson (1977), Hector–Hirsch (1981, 1983), Reinhart (1983), Camacho–Neto (1985), Molino (1988), Tondeur (1988, 1997), Godbillon (1991), Candel–Conlon (2000). Some of these contain detailed historical remarks, or an extensive bibliography. The same is true (although to a smaller extent) for Lie groupoids. There are now various books on Lie groupoids emphasizing different aspects, e.g. Mackenzie (1987), Connes (1994), Cannas da Silva–Weinstein (1999), Paterson (1999); also, a new book by K.C.H. Mackenzie is expected in the near future from Cambridge University Press. The last contains an elaborate history of the subject and an extensive bibliography.

The material in Chapters 1 and 2 is standard, and can be found in some form in many sources, including the books on foliations mentioned above. Let us just mention that the Reeb stability theorems go back to the work of Ehresmann and Reeb in the 1940s; an early reference is Reeb (1952). Our proof of the Reeb–Thurston stability theorems (Section 2.6) follows in part the original proof given by Thurston (1974) and the one given in Mrčun (1996). For more on foliations with compact leaves, see also Epstein (1976). Orbifolds were first introduced under the name V-manifolds in Satake (1956). The notion of holonomy goes back to the work of Ehresmann, Reeb and Haefliger in the 1950s. The notions of Riemannian foliation and bundle-like metric originate with Reinhart (1959).

The classical theorems on codimension 1 foliations presented in Chapter 3 can be found in Haefliger (1956, 1958) and Novikov (1964); see also Rosenberg–Roussarie (1970). There are now simpler proofs of Haefliger's theorem which use the transverse structure and Lie groupoids, see e.g. Jekel (1976) or Van Est (1984).

Much of the material on Riemannian foliations presented in Chapter 4 is due to Molino (1988).

Groupoids have a long history, going back to work of Brandt at the beginning of the twentieth century. The first explicit construction of the holonomy groupoid of a foliation is usually attributed to Winkelnkemper (1983). However, in the study of foliations the use of pseudogroups (which we have seen to be equivalent to effective étale groupoids) is much older; see e.g. Haefliger (1958). Haefliger was also one of the first to use étale groupoids in the study of orbifolds (see Haefliger (1984)); the exact correspondence between orbifolds and proper effective groupoids seems to originate with Moerdijk–Pronk (1997). For more on proper groupoids, see Weinstein (2002).

The notion of Lie algebroid is due to Pradines, who outlined a perfect 'Lie theory' for Lie groupoids and Lie algebroids (Pradines (1967)), parallel to the usual theory for Lie groups and Lie algebras, but without giving any proofs. It was only much later when the theory of Almeida and Molino (1985) showed that the analogue of Lie's third theorem fails for Lie algebroids (cf. Theorem 6.10). Mackenzie's book (1987) contains a cohomological obstruction to the integrability of transitive Lie algebroids. For recent advances on the problem of integrability of Lie algebroids, see Crainic–Fernandes (2003) and references cited there. The paper of Higgins and Mackenzie (1990) discusses some of the categorical aspects of Lie algebroids and Lie groupoids. Our exposition in Section 6.3 of the parts of Lie theory which do easily work for Lie algebroids is based on Moerdijk–Mrčun (2002), where the reader can find references to special cases which were known earlier. Proposition 6.8 also occurs in Mackenzie–Xu (2000).

Almeida, R. and Molino, P. (1985). Suites d'Atiyah et feuilletages transversalement complets, *C. R. Acad. Sci. Paris* **300**, 13–15.

Bott, R. (1972). Lectures on characteristic classes and foliations, in *Lectures on Algebraic and Differential Topology*, editors R. Bott, S. Gitler, I.M. James, Lecture Notes in Mathematics **279**, 1–94 (Springer-Verlag, New York).

Bott, R. and Tu, L.W. (1982). *Differential Forms in Algebraic Topology* (Springer-Verlag, New York).

Camacho, C. and Neto, A. (1985). *Geometric Theory of Foliations* (Birkhäuser, Boston, Massachusetts).

Candel, A. and Conlon, L. (2000). *Foliations I*, Graduate Studies in Mathematics **23** (American Mathematical Society, Providence, Rhode Island).

Cannas da Silva, A. and Weinstein, A. (1999). *Geometric Models for Noncommutative Algebras*, Berkeley Mathematics Lecture Notes **10** (American Mathematical Society, Providence, Rhode Island).

Connes, A. (1994). *Noncommutative Geometry* (Academic Press, San Diego).

Crainic, M. and Fernandes, R.L. (2003). Integrability of Lie brackets, *Ann. Math.* **157**, 575–620.

Dazord, P. (1997). Groupoïde d'holonomie et géométrie globale, *C. R. Acad. Sci. Paris* **324**, 77–80.

Douady, A. and Lazard, M. (1966). Espaces fibrés en algèbres de Lie et en groupes, *Invent. Math.* **1**, 133–151.

Dupont, J.L. (1978). *Curvature and Characteristic Classes*, Lecture Notes in Mathematics **640** (Springer-Verlag, New York).

Epstein, D.B.A. (1976). Foliations with all leaves compact, *Ann. Inst. Fourier (Grenoble)* **26** *no. 1*, 265–282.

Est, W.T. van (1984). Rapport sur les *S*-atlas, *Astérisque* **116**, 235–292.

Godbillon, C. (1991). *Feuilletages. Etudes géométriques*, Progress in Mathematics **98** (Birkhäuser, Basel).

Golubitsky, M. and Guillemin, V. (1973). *Stable Mappings and Their Singularities* (Springer-Verlag, New York).

Greub, W., Halperin, S. and Vanstone, R. (1978). *Connections, Curvature and Cohomology*, Pure and Applied Mathematics: a Series of Monographs and Textbooks **47** Vol. I (Academic Press, New York).

Guillemin, V. and Pollack, A. (1974). *Differential Topology* (Prentice-Hall, Englewood Cliffs, New Jersey).

Haefliger, A. (1956). Sur les feuilletages analytiques, *C. R. Acad. Sci. Paris* **242**, 2908–2910.

Haefliger, A. (1958). Structures feuilletées et cohomologie à valeur dans un faisceau de groupoïdes, *Comment. Math. Helv.* **32**, 248–329.

Haefliger, A. (1984). Groupoïdes d'holonomie et classifiants (with an appendix written with Quach Ngoc Du), *Astérisque* **116**, 70–107.

Haefliger, A. and Reeb, G. (1957). Variétés (non séparées) à une dimension et structures feuilletées du plan, *Enseignement Math. (2)* **3**, 107–125.

Hector, G. and Hirsch, V. (1981, 1983). *Introduction to the Geometry of Foliations Part A, Part B* (Vieweg, Braunschweig).

Higgins, P.J. and Mackenzie, K.C.H. (1990). Algebraic constructions in the category of Lie algebroids, *J. Algebra* **129**, 194–230.

Hirsch, M.W. (1976). *Differential Topology* (Springer–Verlag, New York).

Jekel, S. (1976). On two theorems of A. Haefliger concerning foliations, *Topology* **15**, 267–271.

Kamber, F.W. and Tondeur, Ph. (1975). *Foliated Bundles and Characteristic Classes*, Lecture Notes in Mathematics **493** (Springer-Verlag, New York).

Kobayashi, S. and Nomizu, K. (1963). *Foundations of Differential Geometry* (Interscience publishers, New York).

Lawson, H.B. (1977). *The Quantitative Theory of Foliations*, Regional Conference Series in Mathematics **27** (American Mathematical Society, Providence, Rhode Island).

Mackenzie, K.C.H. (1987). *Lie Groupoids and Lie Algebroids in Differential Geometry*, London Mathematical Society Lecture Notes Series **124** (Cambridge University Press).

Mackenzie, K.C.H. (in press). *The General Theory of Lie Groupoids and Lie Algebroids* (Cambridge University Press).

Mackenzie, K.C.H. and Xu, P. (2000). Integration of Lie bialgebroids, *Topology* **39**, 445–467.

Mac Lane, S. (1963). *Homology*, Grundlehren der mathematischen Wissenschaften **114** (Springer-Verlag, Berlin).

Milnor, J. (1963). *Morse Theory* (Princeton University Press).

Moerdijk, I. and Mrčun, J. (2002). On integrability of infinitesimal actions, *Amer. J. Math.* **124** no. *3*, 567–593.

Moerdijk, I. and Pronk, D. A. (1997). Orbifolds, sheaves and groupoids, *K-Theory* **12** no. *1*, 3–21.

Molino, P. (1988). *Riemannian Foliations* (Birkhäuser, Boston, Massachusetts).

Mrčun, J. (1996). An extension of the Reeb stability theorem, *Topology Appl.* **70**, 25–55.

Novikov, S. P. (1964). Foliations of co-dimension 1 on manifolds (in Russian), *Dokl. Akad. Nauk SSSR* **155**, 1010–1013.

Paterson, A.L.T. (1999). *Groupoids, Inverse Semigroups, and Their Operator Algebras*, Progress in Mathematics **170** (Birkhäuser, Boston, Massachusetts).

Pradines, J. (1967). Théorie de Lie pour les groupoïdes différentiables, calcul différentiel dans la catégorie des groupoïdes infinitésimaux, *C. R. Acad. Sci. Paris* **264**, 245–248.

Reeb, G. (1952). Sur certaines propriétés topologiques des variétés feuilletées, *Actual. Sci. Ind.* **1183** (Hermann, Paris).

Reinhart, B.L. (1959). Foliated manifolds with bundle-like metrics, *Ann. Math. (2)* **69**, 119–132.

Reinhart, B.L. (1983). *Differential Geometry of Foliations. The Fundamental Integrability Problem*, Ergebnisse der Mathematik und ihrer Grenzgebiete **99** (Springer-Verlag, Berlin).

Rosenberg, H. and Roussarie, R. (1970). Reeb foliations, *Ann. Math.* **91**, 1–24.

Satake, I. (1956). On a generalization of the notion of manifold, *Proc. Nat. Acad. Sci. U.S.A.* **42**, 359–363.

Serre, J.-P. (1965). *Lie Algebras and Lie Groups* (W.A. Benjamin, New York).

Thurston, W.P. (1974). A generalization of the Reeb stability theorem, *Topology* **13**, 347–352.

Tondeur, Ph. (1988). *Foliations on Riemannian Manifolds* (Springer-Verlag, Berlin).

Tondeur, Ph. (1997). *Geometry of Foliations*, Monographs in Mathematics **90** (Birkhäuser, Basel).

Warner, F.W. (1983). *Foundations of Differentiable Manifolds and Lie Groups* (Springer-Verlag, New York).

Weinstein, A. (2002). Linearization of regular proper groupoids, *J. Inst. Math. Jussieu* **1** no. *3*, 493–511.

Winkelnkemper, H.E. (1983). The graph of a foliation, *Ann. Global Anal. Geom.* **1** no. *3*, 51–75.

Index

Printed in the United States
By Bookmasters